THE BEST REMAINING SEATS

THE GOLDEN AGE OF THE MOVIE PALACE

BY BEN M. HALL

Foreword by Bosley Crowther
New preface by B. Andrew Corsini

A Da Capo Paperback

ACKNOWLEDGMENTS

The author is deeply grateful to Warren Stokes who designed this book, and — as if that were not enough, was counselor, goad and moral supporter all through the years of its preparation; to Hugh Rogers who took many of the photographs which illustrate the text; to Arthur Kloth who indexed it; to Martin Goldblatt who pored over it; and to Lenore and David Redstone who served as copy editors. Special thanks go to Brother Andrew Corsini, advocate of the impossible.

For furnishing recollections, photographs and advice, as well as opening doors, granting permissions and being generally kind, the author bows to the following: Mary Squire Abbot, Noble Arnold, Leon Berry, Clealan Blakely, Eugene Braun, Frank Cambria, Jesse Crawford, Regina Crewe, Mel Doner, Drew Eberson, Ernest Emerling, Tom Felice, Maria Gambarelli, Marjorie Geiss, Chuck Gerhardt, Paula Gould, Ben Grauer, Chester Hale, Reginald Hoidge, John Jackson, Eva Kirchmeier, Arthur Knorr, Charles Kurtzman, John Landon, Dr. Ray Lawson, Ann Leaf, Leon Leonidoff, Mrs. Lawrence Levy, Harold Lloyd, Richard Loderhose, Helen Lupo, Arthur Mayer, Carmel Meyers, Allen Miller, Roger Mumbrue, Leif Neandross, Daniel Papp, C. A. J. Parmentier, Jack Partington, Jr., E. J. Quinby, Harold W. Rambusch, Henry Hope Reed, Jr., Robert C. Rothafel, Miles Rudisill, Jr., Ben Selvin, Edward Serlin, Raymond Shelley, Nathaniel Shilkret, Richard Simonton, Barry Spencer, Douglas Stanbury, Mildred Strelitz, Gloria Swanson, A. E. Tovey, Julia Fay Treloar, Carlton Winckler, Robert Youngson, and William Zeckendorf.

Special acknowledgment is extended to Brown Brothers, Culver Service, Eliot Elisofon and *Life*, Manning Brothers, Edgar Orr, the Seattle *Times* and particularly Terry Helgesen, for photographs; to Paul Myers, and the Theatre Collection, New York Public Library; to George Schutz, Associate Editor of *Motion Picture Herald;* to Winfield Andrus, Managing Editor of *Film Daily;* to Harold Hunt of The Spot and Marc Ricci of The Memory Shop; to the American Heritage Publishing Co., for use of color plates; and to the editors of *Variety, The New Yorker, The New York Times,* and *Billboard,* for indispensable reference material.

Library of Congress Cataloging in Publication Data

Hall, Ben M.
The best remaining seats.

(A Da Capo paperback)
Rev. ed. of: The golden age of the movie palace. 1975, c1961.
Includes index.
1. Motion picture theaters–United States–History. 2. Theaters–United States–Construction. I. Hall, Ben M. Golden age of the movie palace. II. Title.
PN199.3.5.U6H27 1988 791.43′0973 88-288
ISBN 0-306-80315-1

This Da Capo Press paperback edition of *The Best Remaining Seats* is a republication of the revised edition published in New York in 1975 and issued under the title *The Golden Age of the Movie Palace.* The present edition has been updated through 1987 and is supplemented with a new preface by B. Andrew Corsini. It is reprinted by arrangement with Crown Publishers, Inc.

Published by Da Capo Press, Inc.
A Subsidiary of Plenum Publishing Corporation
233 Spring Street, New York, N.Y. 10013

Manufactured in the United States of America

Preface

Certainly both Ben Hall and Bosley Crowther, who wrote the Foreword, were honestly pessimistic in their outlook for the future of the motion picture theatre when this book was originally published in 1961. It is an understandable attitude, given the events of the time. The large movie palaces, or "cathedrals" as Crowther refers to them, were certainly doomed; many had already been lost and the foreseeable future indeed looked gloomy.

While the chosen title of *The Best Remaining Seats* does appear to confirm that the author was hinting at the soon-to-come nation-wide upheaval within the field of motion picture exhibition, there was also, perhaps unknowingly, a forecast of the future for at least a portion of those fabulous movie palaces. For out of the reality of the destruction, during this early period, of some of the major examples of theatre architecture, a totally surprising and unforeseen force arose: a widespread movement to save some prime examples of this specialized form of architecture. This force caught hold all across the United States and continues to this day.

Theatres in both large and small cities have been purchased by groups of civic-minded citizens, as well as some individuals, to be converted into "performing arts centers," to serve as "road" houses for touring productions, or as "home" for a local symphony orchestra or opera company. The medium and smaller-sized cities have led the parade in this endeavor. Strangely enough, few of the larger cities can be included in this group aside from Cleveland, St. Louis, Chicago, and Boston.

From the first edition of this book came the incentive to save some of these theatres and keep them alive for future generations of young people; also born from this effort was the Theatre Historical Society of America. The Society was founded by Ben Hall in 1969 with less than 100 founding members, all of whom Ben knew personally. Today the Society has grown to eight times that number and maintains a Service Center in Chicago housing a large archives for research purposes, both for the society's members and for interested scholars. These archives have served scores of renewal and restoration projects over the years.

So thanks to Ben Hall many of those "Best Remaining Seats" do indeed remain. Any and all persons interested in architectural history or entertainment memorabilia owe a large debt of gratitude to Mr. Hall for the forces he set in motion, and he himself would be delightedly astonished as a child finding the real Santa Claus.

B. Andrew Corsini
September 1987

CONTENTS

FOREWORD by Bosley Crowther

IN ALL of the many learned histories of motion pictures that have been published over the years, surprisingly little close attention has been given to the places in which those pictures have been shown.

To be sure, there has been some wistful interest in the nickelodeons as the ingeniously adapted habitations of the infant silent films, and some of the shrewder historians have marked the significance in the rise and successive elaborations of the great movie palaces. But no one — at least, to my knowledge — has heretofore done a straight-out book on theatre architecture and its relation to the exhibition of films.

This is the more surprising because the fact should be recognized that the total effect of a motion picture is conditioned to a greater or lesser extent by the environment in which it is shown. Run off a film of a certain character in a modest screening room and it curiously lacks the magnetism it might have in a well-filled theatre. Show it at an open-air drive-in and its effect may be something else again. The "chemistry" (as they now call it) of its emanations is subtly changed by the surrounding atmosphere.

Further, the fun of movie-going, the pleasure that is derived from the experience of spending a few hours in a movie theatre, includes a lot more than the experience of simply observing the film. It is the amalgamation of a series of pleasurable stimuli — the initial anticipation, the warmth of companionship, the congeniality of the surroundings, the freedom to use and partake of the facilities of the auditorium, the feeling of elegance. The extent of one's final satisfaction is in the *total experience.*

Thus it should well behoove historians to pay more attention to the physical potentialities of enjoyment provided by theatres.

For this reason, we should be grateful to Ben M. Hall for giving us this book that makes an appropriate exploration into this area. For here at last is a volume which properly and feelingly surveys that most lofty and awesome endeavor of the theatre builders' design, the extravagant Movie Palace — or, better, Cathedral.

Those of us who have lived through the conspicuously brief but gaudy age of these cultural manifestations and have witnessed the weird and wild conceits of the latter-day temple builders and the high priests of the rituals held therein should find in these vivid recollections a goad to our own fond memories, as well as a glittering crystallization of colorful history. Those who are too young for remembering but have a liberal credulity should find them a source of amazement and appropriately fanciful delight.

Inevitably, in perusing this phantasmagoric account, which is rendered the more entertaining (and convincing!) by the stunning photographs, the reader is going to do some wondering about the delicate matter of taste and the question of the aesthetic standards of the theatre architects.

Let us not be too hoity-toity with our judgments in this regard. The intent of the temple builders and the wizards who elaborately conceived the stage shows and other bold attractions that adorned these theatres, was not to please Lewis Mumford and serious critics of American art. It was to attract the susceptible mass audience and to delight it with extreme, eye-filling shows.

And no one who has done much sharp observing of the American scene is likely to condemn these fabricators for misjudging massive public taste. One needs but consider the monstrosities of architecture and *décor* that our people have happily tolerated (some of them well-placed people, too) in the time that our overgrown cities and social concentrations have emerged to recognize that the movie meccas were in no way unique. Look at the frightening decorations of Victorian-era "gingerbread," the Spanish and Moorish concoctions in resort and suburban homes, the razzle-dazzle of modern supermarkets and the absurdities of motor cars with fins. We are a people who go for — and have gone for — the gaudy and bizarre.

The motion-picture palaces were, in their time and prime, the peculiar satisfiers to a massive public taste. This is a realization that Mr. Hall most felicitously conveys. Just as the run of the pictures that were the basic elements of their programs — were, indeed, the inspirations for their structure and their continuing *raisons d'etre* — provided the sensation-hungry with illusions and fantasies, the great movie theatres did the same thing with their size and munificence. They were the tangible illusions in which the more shadowy illusions were contained. Thus they contributed greatly to a *total experience* of escape.

So let us not be too impatient with the weirdly inventive architects who dreamed up these cinematic temples or with the craftsmen who made them what they were. In the annals of American culture, the names of John Eberson and Thomas Lamb should be as notable and potent as the names of Frank Lloyd Wright and Ralph Adams Cram, the name of S. L. Rothafel — or Roxy — as considerable as that of Eugene O'Neill.

Perhaps you have by now detected that I am studiously employing the past tense. I am because the huge movie palace is now pretty much a thing of the past. Even before its fullest flowering (as Mr. Hall explains), the factors that made for its fading were coming into play.

The introduction of sound motion pictures brought a certain realism to the screen that changed the quality of illusion provided by the silent film. Music, talk and natural noises were not entirely in accord with the lavish unreality of the theatres or the pastiche nature of their lush stage shows.

But much more disastrous to these theatres than this minor aesthetic clash was the loss of the mass movie audience to the new home television device. With the mammoth expansion of that medium, American movie-going declined and the huge multi-thousand-seat theatres began to slip into the red. The demolition (or alteration) of such one-time prosperous picture palaces as the Mastbaum in Philadelphia, the Roxy in New York, the Midland in Kansas City and dozens of others all over the land is solemn testimony to the passing of an age.

Nowadays it is the small, compact "art houses" and once-big theatres quickly redesigned into smaller and sleeker auditoria that are drawing the bulk of the urban movie trade with generally sophisticated or spectacularly graphic films. As yet, a thoroughly novel style of theatre for the new conditions has not been designed — or, if it has, in some experimental cases, it hasn't got the widespread confidence of financiers.

Meanwhile, the last of the old "cathedrals" (with the exception of the Radio City Music Hall, which remains a uniquely prosperous instance of a high style, modern presentation house) stand, either dead or slowly dying, memorials to a time when the public was more devout in its movie-going.

This book is their appropriate epitaph.

"The Cathedral of the Motion Picture"

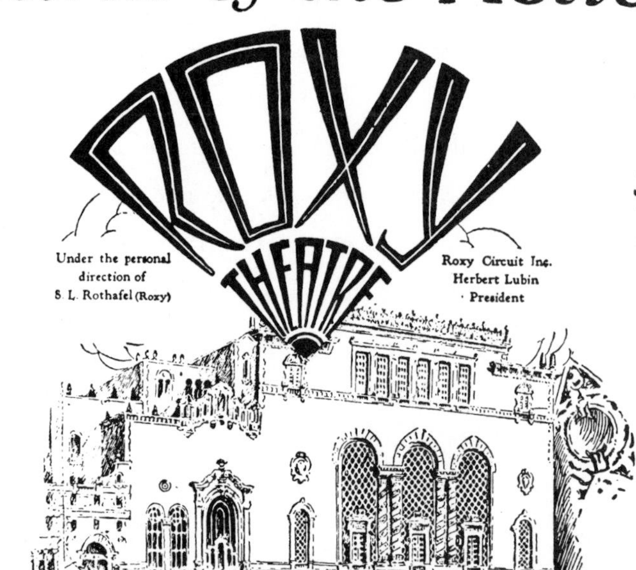

WHAT SHALL WE SAY?

We cannot find adjectives and superlatives strong enough to describe the thousand and one wonders and innovations of The ROXY, truly the most sumptuous and stupendous theatre ever erected.

The ROXY will give you an unforgettable thrill. In all the world there is no theatre like it.

facts

about the $10,000,000 ROXY THEATRE

- World's largest theatre — seats over 6,200.
- Foyers and lobbies of incomparable size and splendor.
- Decorations of indescribable beauty.
- A new idea in stage and stagecraft.
- Acoustics — *A revelation!*
- Projection — *Another revelation.*
- Ventilation: The last word in scientific air-conditioning.
- Spacious elevators to the balcony.
- Lighting: a plant three times the capacity of any other theatre—sufficient to light and power a city of a quarter of a million.
- Luxurious and comfortable seats, arranged to provide *unusually* ample room between rows.
- Six box offices conveniently located for your service.
- Service: A staff of attendants thoroughly organized and drilled under the direction of a retired Colonel of the U. S. Marines, ensures every courtesy.
- *Standards of entertainment never before attempted.*
- *Largest permanent symphony orchestra in existence.*
- *Colossal pipe organ—largest in any theatre in the world—played simultaneously by three organists on three separate consoles.*
- *Permanent choral group of 100 voices.*
- *Permanent ballet corps of 50 dancers.*
- *Cathedral chimes of 21 bells—first time in any theatre.*
- *and ... The VITAPHONE!!!*

Opening with
GLORIA SWANSON'S
Greatest Dramatic Triumph
"The Love of Sunya"
Her First United Artists Production

ROXY THEATRE 50th ST. & 7th AVE.

Premier FRIDAY MARCH 11th at 8:30 P.M.

At a moment shortly after eight o'clock on March 11, 1927, the gaudy, enchanted, phony, preposterous, and lovely Golden Age of the Movie Palace reached its Klieg-lit pinnacle. Nothing quite like it was ever to happen again...

HOTEL MANGER
WIDE
WIDE

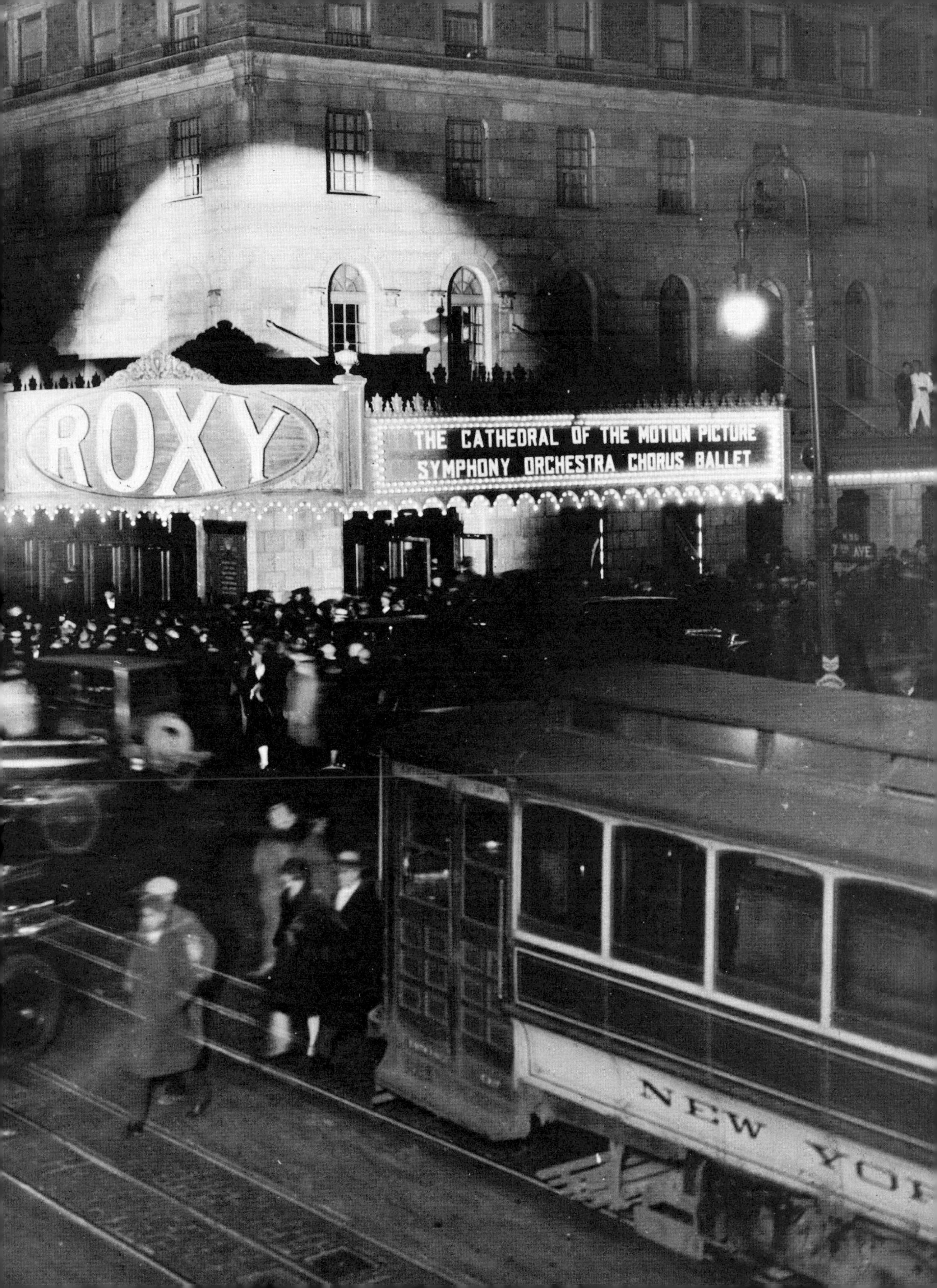

ROXY
THE CATHEDRAL OF THE MOTION PICTURE
SYMPHONY ORCHESTRA CHORUS BALLET
7TH AVE
NEW YOR

Even Texas Guinan was there....

Since midafternoon they had been gathering. By seven o'clock the intersection of Seventh Avenue and Fiftieth Street in New York was jammed with nearly ten thousand people, and more were arriving on every trolley car and subway.

Deputy Inspector McGrath, veteran of many a New Year's Eve in Times Square, felt right at home. He stood in the tonneau of a black Lincoln touring car, shouting orders to the hundred and twenty-five cops in his command as they struggled to keep the crowd behind the barricades. Reinforcements were on the way from the West Forty-seventh Street station house; McGrath hoped they would get there soon.

The raw March wind nipped the ankles of the flappers and made the loose latches of their galoshes jingle; housewives pulled their cloches down tighter over skimpy boyish bobs, and their husbands stamped their feet and were thankful they still had on their winter BVDs. They were an orderly, good-natured crowd, but as the arc lights came on and swept back and forth over the scene, their excitement became something that McGrath, with his cop's awareness of such things, could feel on the back of his neck.

Soon the procession of Marmons, Minervas, and Pierce-Arrows began to join the Checker cabs, plowing through the crowd to discharge their passengers under the marquee. As the first celebrities arrived — Richard Dix or Lois Wilson, or possibly Mary Brian or Ben Bernie—the barricades went down like jackstraws, leaving McGrath and his Finest to operate as skirmishers.

It wasn't long before arrival at the Roxy's doors proved to be both perilous and impractical. Mrs. Otto Kahn, her limousine foundered on the shoals of Roseland, made a valiant march to the theatre in the wake of chauffeur and footman. Harold Lloyd, coming from the opposite direction, was recognized under the marquee even without his

Celebrities Flock to Brilliant Opening of Roxy's Movie Cathedral

(Photo Graphic.)

BRILLIANT audience of 6,200 leaders in city's commercial, professional and artistic life filled Roxy Theater at opening of the "Cathedral of the Movie" last night. Photo above shows part of throng that filled theater. (Story on Page 2)

(Photo Graphic.)

GOING TO THE MOVIES. Our mayor, Jimm[y] Walker, always th[e] last word in style, dolled up especially smart for Roxy The[-] ater opening. He's shown, accompanied by Mrs. Walke[r,] entering magnificent new cinema palace.

"THE VIRGIN WIFE" of reel life was one of the guests at Roxy opening last night. In real life, Pauline Garon (above), who has completed work in True Story picture version of "The Virgin Wife," is wife of Lowell Sherman, stage and screen star.

(Photo Graphic.)

CELEBRITY. That's the word that fits nearly all the guests on hand for the Roxy Theater opening. Photo above shows Alice Joyce, screen star, who was an early arrival.

(NEWS photo)

ROXY'S MOVIE CATHEDRAL OPENS!—Thousands last night jammed street at opening of Roxy Rothafel's movie palace, which seats 6,200 persons. Here are Louise Davis, Roxy, Mrs. Rothafel, Harold Lloyd and Roxy's daughter, Anne (l. to r.), snapped during the ceremonies—Story, p. 4.

INDEPENDENT EXHIBIT OPENS. Modern themes, modernistically handled, feature Independent art show that opened last night at the Waldorf. Photo shows Mme. Anna de Gombert with her plaque of the late Rudolph Valentino.

horn-rimmed glasses; ten minutes later he and his wife struggled into the safety of the lobby, visibly shaken — and impressed. Hope Hampton, even then an old hand at opening nights, feared for the safety of her diamonds and clung to her husband, Jules Brulatour, as he cleared a path with his cane.

None of the good-natured mob that surged over the sidewalks with cries of "Hey, Charlie . . . where's your mustache?" and "It's Norma Talmadge!" had any real hope of getting into the theatre that night. Those who held the special green opening-night tickets had bought them (at eleven dollars apiece) well in advance. The rest—and they made up almost the entire population of the orchestra floor—arrived bearing engraved invitations.

The members of the press got in by showing the gold pencils (inscribed with their names) which had been issued them as lifetime passes.

Mr. and Mrs. H. W. Llewellyn, of Newark, N. J., got to the Roxy early. The Llewellyns had had the foresight in November 1925 to send a blank check requesting two seats for opening night. Now, a year and a half later and bearing engraved cards instead of tickets, they stood at the bronze doors of the Cathedral of the Motion Picture, blinking in the glare of the floodlights. Then, as a barrage of photographers' flash powder went off and the crowd began to shout, the Llewellyns were swept inside on a wave of spangles, spit curls, and opera hats.

For a while they bobbed around in the Rotunda, a little floating island of delighted disbelief, until finally they were swallowed up in the dim vastness of the auditorium, to be seen no more.

•

The original plan for opening night had called for the Roxy uniformed staff to guide the audience in groups on a tour of the theatre during the hour before the curtain went up. But this idea evaporated in confusion as the visitors organized their own sight-seeing tours. Every ornamental niche, balcony, and winding staircase was jammed with goggle-eyed first-nighters. Ushers, standing guard at the footlights, protected the mysteries of the sanctum sanctorum, but elsewhere in the Cathedral of the Motion Picture nothing was sacred, as bands of women made giggling forays into the paneled Tudor stronghold of the Gentlemen's Smoking Lounge, and groups of gentlemen returned the compliment elsewhere.

At least three thousand of the visitors were content just to stand in the Rotunda outside the auditorium, gaping. They marveled at the great crystal chandelier and the five-story columns of *verde* marble, and sank to their shoe-tops in the Largest Oval Rug in the World. During lulls in the excitement, they could hear the music of the Rotunda's pipe organ floating out from the Musicians' Gallery over the oohs and ahs of the crowd.

As they slowly pushed and were pushed into the auditorium, most of them got their first glimpse of the man in whose name all this splendor had come to pass: Samuel Lionel Rothafel — Roxy himself. He stood beneath an arrangement of red and white carnations that spelled out R-O-X-Y . . . a poker-faced Buddha in a tuxedo whose smile was turned on almost as if a button had been pressed whenever he saw a friend in the crowd. Standing beside Roxy was his wife, Rosa, a shy, kind-looking woman in a velvet cloak trimmed with white fur. With them was their daughter, Beta, a pretty and vivacious girl with an outdoor freshness that seemed slightly out of place in that crowd. The air around the Rothafels was heavy with the smoke of flash powder as each movie idol or captain of industry who entered paused to be photographed with Roxy. Just as the smoke had begun to clear and it seemed as if the Rothafels' ordeal by flashlight might be over, there came a roar from outside and the crowd in the Rotunda parted to form a path across the oval rug. In swept the Marquise de la Falaise de la Coudraye, her astonishing white teeth and patent-leather coiffure glistening as the lights flashed again and again. Following in her wake was the Marquis—quite content to be just plain Mr. Gloria Swanson for the evening.

Gloria stood beside Roxy and he kissed both her hands. It was a night for her to remember.

The premiere attraction on the Roxy screen that night was to be *The Love of Sunya,* starring Gloria Swanson and produced by Gloria Swanson, who had, at last, become one of the United Artists.

By eight-thirty nearly everyone was settled in the 6,214 seats (whose red-plush backs all bore the ubiquitous "R" monogram). As nine o'clock neared, only two seats were still vacant down front, but everybody seemed to understand about those.

Suddenly, from high up on the left, above the stage, three chime notes pealed out and a hush settled over the program rustlers below. The amber house lights dimmed. Majestically, from the bowels of the orchestra pit, rose not one, not two, but three great golden organ consoles, each manned by an organist in a sort of green velvet smoking jacket. Dezso Von D'Antalffy, Emil Velazco, and Casimir A. J. Parmentier were their names and they thundered through "The Pilgrims' Chorus" with a fervor befitting the occasion. As they throbbed into "Londonderry Air," there came a distraction: a flutter of applause broke out at the rear of the auditorium and seemed to grow stronger as it washed down toward the orchestra pit.

The organists kept bravely about their business as the demonstration grew louder, and if they turned around to peek, it is to their everlasting credit that they didn't swing into "Will You Love Me in December?"—the anthem usually required for such an occasion. For, late as usual, Mayor James J. Walker had arrived. And this time, for a change, the lady with him was Mrs. Walker.

The organists sank slowly out of sight with what dignity they had left. Once again the chimes rang out, and an ecclesiastical stillness fell over the Cathedral of the Motion Picture. On the third note, every light in the place went out. Even the exit signs.

Slowly a baby spot picked out the head and shoulders of a figure robed as a monk, somewhere high and far away. From a scroll, and in a voice full of portent, he read the Invocation:

> Ye portals bright, high and majestic, open to our gaze the path to Wonderland, and show us the realm where fantasy reigns, where romance, where adventure flourish. Let ev'ry day's toil be forgotten under thy sheltering roof —O glorious, mighty hall—thy magic and thy charm unite us all to worship at beauty's throne . . .
>
> Let there be light.

And there was light.

Great golden floods of it burst over the orchestra pit (which only moments before had been empty and seemingly bottomless) to reveal the

Roxy Symphony Orchestra of a hundred-ten musicians and four conductors: Erno Rapee, Charles Previn, H. Maurice Jacquet, and Frederik Stahlberg. When the applause for this miracle had died down, Conductor Jacquet rapped for silence and raised his baton for the overture.

It was billed on the program as a symphonic tone poem, a musical description of the events surrounding the writing of "The Star Spangled Banner." As the overture progressed, the gold plush curtains opened for the first time, and there was the dawn sky over Fort McHenry, lit by flashes of artillery fire. Cannons boomed, a chorus in somber garb (which had suddenly appeared from nowhere) grouped around Francis Scott Key. Reading over his shoulder, they swung into the refrain he was feverishly scribbling as he glimpsed that the flag was still there.

Next day, *The New York Times* described the finale this way: "It was while Roxy was leaning over the rail of a ship bound for Europe that he obtained the idea for this scene . . . a burnt-orange sunrise with the stars just visible. This was gradually transformed, through streaky clouds, into the American Flag. It was accomplished with marked artistry, and the audience arose as the orchestra played the National Anthem."

Having drawn from Genesis and Francis Scott Key, Roxy next borrowed from Nature with "A Floral Fantasy." This was choreographed by Leo Staats, *maître de ballet* (formerly of the Paris Opera House), assisted by Leon Leonidoff, and proved to be a showcase for Maria Gambarelli, Roxy's beloved prima ballerina. Mlle. Gambarelli (or "Gamby," as she was known in Radioland to the millions of fans of the "Roxy's Gang" programs — she of the irrepressible giggle) danced The Fairy in the ballet. A great burst of applause greeted Gamby when she first appeared from behind the enormous *art nouveau* tree that dominated the stage. Brave little trouper that she was, she managed to save the entire *corps de ballet* (representing a dozen Roses, a dozen Carnations, and a dozen Sylphs) from the insidious advances of Allan Wayne as The Wind, Harold Ames as The Storm, and Alex Fisher as Lightning.

Next on the bill was the screening of specially filmed greetings (seen but not heard) from President Coolidge, Mayor Walker, Governor Smith, Vice-President Dawes, and Thomas A. Edison. This was climaxed by a scene showing three hundred patients at Walter Reed Hospital arranged on the lawn to spell out Roxy's name, and it was announced that the proceeds from the opening night would be used to buy radios for the patients in veterans' hospitals all over the country.

Clips from Fox newsreels chronicled the Roxy's rise from first steelwork to elaborate terra cotta façade. Maria Gambarelli and Gladys Rice and gentlemen of Roxy's Gang report for rehearsal (below). On opening night floodlights play on the marquee and on fur-bearing celebrities (blurred figure belongs to Gloria Swanson).

When the curtains opened again, it was for "A Fantasy of the South." This began with Julius Bledsoe singing "Swanee River" before a setting representing the Jersey waterfront. Obviously his heart was pining for the old folks at home, as revealed in the next scene. Here, the Roxy company took their stand for Dixie in the "Southern Rhapsody" number. Soloists were Florence Mulholland, contralto; J. Parker "Daddy Jim" Coombs, basso; José Santiago, baritone; Margaret "Mickey" McKee, whistler; Charlotte Ayers, solo dancer; and the Roxy Chorus and Singing Ensemble.

The scenic effects for this spectacle — from the rippling waters off Hoboken to the stately white columns and possums buzzing in the cotton vines — were projected for the first time in any theatre by the Trans-Lux system on a translucent screen with projectors at the rear.

This was followed by the Roxy Pictorial Review, a newsreel put together from the best shots of all the commercial newsreels of the week. Fox News stole the applause here with a documentary on Roxy, The Gang, and the construction of the theatre as seen in various stages.

The eighth item on the program was "A Russian Lullaby." Irving Berlin had been specially commissioned to write the song for the opening of the Roxy. Douglas Stanbury, stalwart baritone of Roxy's Gang, stood on the stage apron, uniformed as a cossack. As he sang, the lights behind a scrim curtain came up to reveal Gladys Rice crouched over a cradle, in a hut many wolf-cries away, while an almost invisible chorus rose on an elevator platform at the rear of the stage, humming an accompaniment to the ballad. For this number Frederik Stahlberg conducted the Roxy Symphony.

Erno Rapee—his right arm in a sling from too much rehearsal all the week before—took over the baton to lead the Symphony in a left-handed version of the overture to *Carmen.* As this was going on, the picture sheet was lowered for the Vitaphone Presentation. In this novelty, Giovanni Martinelli and Jeanne Gordon were seen *and heard* on the screen singing airs from the second act of the Bizet opera, accompanied by the Metropolitan Opera chorus, ballet, and orchestra. The new miracle was rather coolly received . . . who wanted singing ghosts with so much living music at hand?

At last came the moment Gloria Swanson had been waiting for. Henry Stevens read a short prologue, and the curtains opened on the screen. Two surprises this time: instead of being edged in the usual black, the Roxy's screen seemed to be floating in a luminous mist. And instead of the usual title, credits, and so forth, Roxy started *The Love of Sunya* with the first scene. It was an elaborate production, full of flashbacks within flashbacks—ancient Egypt, imperial Rome, France before the Revolution. There was much crystal-gazing, a number of reincarnations, and miles of impressionism (hands forming geometric patterns, arty lighting) to distinguish the direction of Albert Parker. And the film introduced a new matinee idol to the screen: John Boles. But it was Gloria's film from beginning to end. It was *her* kind of picture, and she played Sunya with every Swansonism at her command.

It was past midnight when the groggy first-nighters rose to stumble into the ambush of flash powder and flood lights that awaited them outside. One group remained inside the auditorium: Harold Lloyd's wife, Mildred Davis, had lost a diamond and ruby bracelet in the midst of all the applause, and a systematic search in the vicinity of Mrs. Lloyd's seat was being conducted by a platoon of Roxy's gentlemen-Leathernecks. The bracelet was found in some sawdust left behind by carpenters the afternoon before, and after the ushers' "no-gratuities" pledge had been put to its first severe test (and come through with flying colors) the Lloyds left the theatre to face the mob that was only then beginning to break up under the marquee.

As the frankincense seeped from the air-conditioning ducts and the acolytes lowered the bare-bulbed work lamps onto the stage to clean up for the next day's show, Roxy stood with his family on a little balcony looking down into the Rotunda, watching the departing magi leave his new Cathedral.

"I'm happy," he sighed. "Take a look at this stupendous theatre. It's the Roxy and I'm Roxy. I'd rather be Roxy than John D. Rockefeller or Henry Ford."

©
·1925·
Bushong
Worcester
Mass

MAMA, MIDAS, AND MAIN STREET

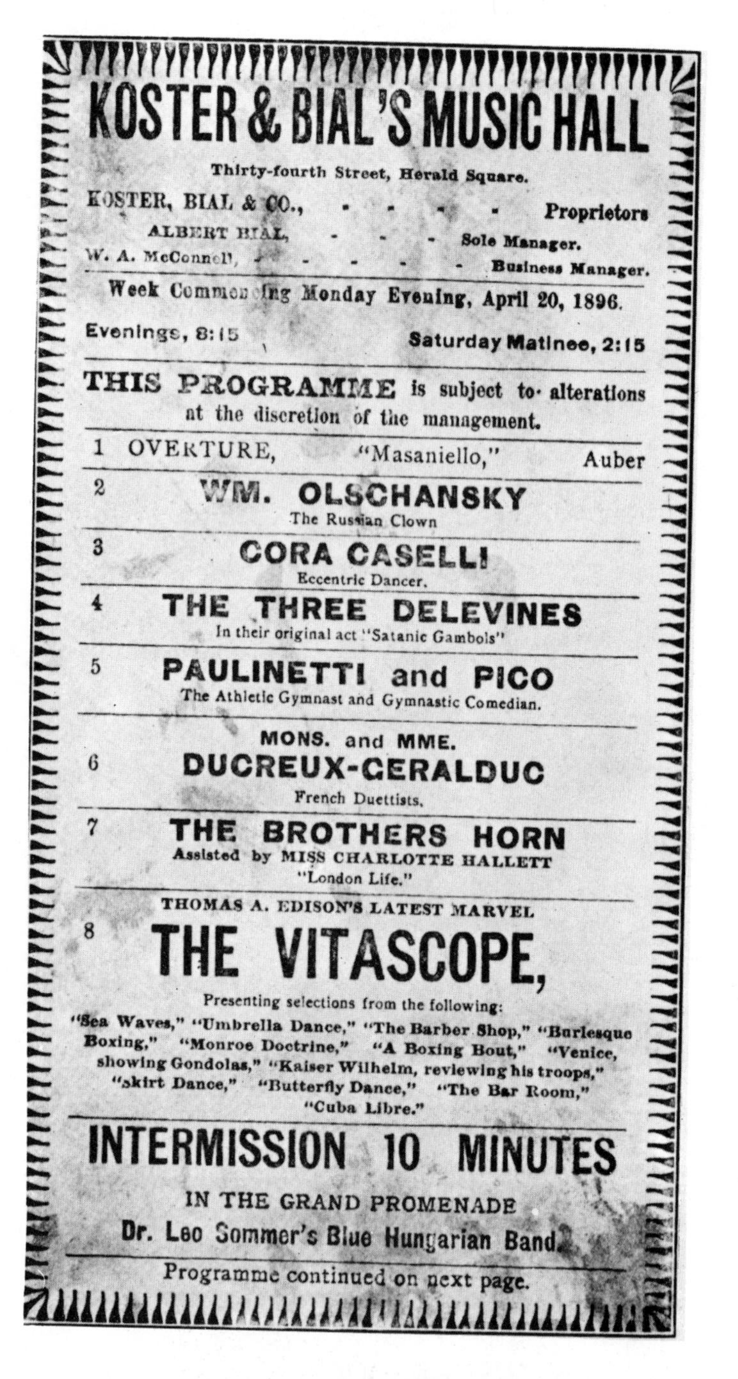

KOSTER & BIAL'S MUSIC HALL

Thirty-fourth Street, Herald Square.

KOSTER, BIAL & CO., ... Proprietors
ALBERT BIAL, ... Sole Manager.
W. A. McConnell, ... Business Manager.

Week Commencing Monday Evening, April 20, 1896.

Evenings, 8:15 Saturday Matinee, 2:15

THIS PROGRAMME is subject to alterations at the discretion of the management.

1 OVERTURE, "Masaniello," Auber

2 WM. OLSCHANSKY
The Russian Clown

3 CORA CASELLI
Eccentric Dancer.

4 THE THREE DELEVINES
In their original act "Satanic Gambols"

5 PAULINETTI and PICO
The Athletic Gymnast and Gymnastic Comedian.

6 MONS. and MME.
DUCREUX-GERALDUC
French Duettists.

7 THE BROTHERS HORN
Assisted by MISS CHARLOTTE HALLETT
"London Life."

8 THOMAS A. EDISON'S LATEST MARVEL
THE VITASCOPE,
Presenting selections from the following:
"Sea Waves," "Umbrella Dance," "The Barber Shop," "Burlesque Boxing," "Monroe Doctrine," "A Boxing Bout," "Venice, showing Gondolas," "Kaiser Wilhelm, reviewing his troops," "Skirt Dance," "Butterfly Dance," "The Bar Room," "Cuba Libre."

INTERMISSION 10 MINUTES
IN THE GRAND PROMENADE
Dr. Leo Sommer's Blue Hungarian Band.

Programme continued on next page.

Though billed for April 20, 1896, the Vitascope didn't appear until the 23rd – world's first postponed movie premiere. Edison ignored "vociferous cheers" and let his "film phantoms" take the bows.

"When you enter these portals you stray magically from the dull world of confusion and cares into a fairy palace whose presiding genius entertains you royally."

— Program note, Capitol Theatre, New York, 1921

And so it was that the golden age of the movie palace reached its zenith on a March night nearly thirty-five years ago.

It was a brief era, as golden ages go. It had swept in on a flood tide of splendor, of million-dollar real estate deals, of fantastic architecture, of music, laughter, and dreams, less than a decade before. And it was to end (along with so many other things) in little more than three years after Roxy consecrated his Cathedral. The whole dizzy, prodigal, enchanted business came to gaudy full bloom, filled the night with its scent, wilted, and drooped in the short span of years that lay between the coming of Prohibition and the onset of the Depression.

The years before the golden age began had been swift and busy and exciting ones for the movies, for the people who showed them, and for the people who were beginning to make them part of their lives.

The U. S. was still on the innocent side of 1910 when it became apparent that movies had come to stay. Traveling tent shows, like Thomas Talley's Electric Theatre, had introduced the miracle of *The Great Train Robbery* to the hinterlands. *Hale's Tours*, an elaborately-conceived movie show that took place in a facsimile railroad coach which rocked and swayed as travelogues were shown at one end to the accompaniment of whistles, bells and hissing steam — was the sensation of the St. Louis Exposition of 1904. Later, under the management of its originator, Chief George C. Hale of the Kansas City Fire Department, *Hale's Tours* toured the world. In that same year — 1904 — Harry, Jack, Albert, and Sam Warner opened a ninety-six-seat store show called The Pioneer in Newcastle, Pa. . . . the Mark brothers, Mitchell and Moe, were busy with their Edisonia Hall in

Buffalo . . . David Grauman and his bushy-haired young son, Sid, were showing movies to small but enthusiastic crowds in a converted store, named the Unique, in San Francisco . . . and John P. Harris, manager of a *musée* and curio hall in McKeesport, Pa., decided to exhibit "living pictures" on an 8 A.M.-to-midnight policy to enthusiastic locals. Harris minted a brand-new name for his movie show; he merged his admission price with the Greek word for theatre and came up with "Nickel-Odeon."

By 1905 nearly any city worth five cents had one or more nickelodeons (alas, Harris's inspired name had become generic with the speed of a pratfall), and legions of Bijoux, Gems, and Cameos were arching their twinkling tungsten façades all across the land.

Only ten quick years had elapsed since the night that Thomas Edison unveiled the Vitascope and projected a program of moving pictures on the gold-framed screen of Koster & Bials' Music Hall on New York's Herald Square (where Macy's now stands). On that historic evening—April 23, 1896—the private imp in the viewing lens of the hand-cranked Kinetoscope leaped like a prancing giant onto the screen, for all to see. From that moment, the movies belonged to the masses.

By 1910, movies were everywhere, and a few entertainment "extras" were coming on the scene. In Milledgeville, Georgia, a show-struck fat boy opened a nickelodeon called the Crescent. With a flair for public service way ahead of his time, he flashed the results of the Johnson-Jeffries world heavyweight championship bout—fought on the afternoon of July 4, 1910, in Reno—that very night on the screen (through a special arrangement with a friend in the local telegraph office). In a few years the fat boy was appearing on the Crescent screen himself in Lubin comedies. His name: Oliver Norvell Hardy.

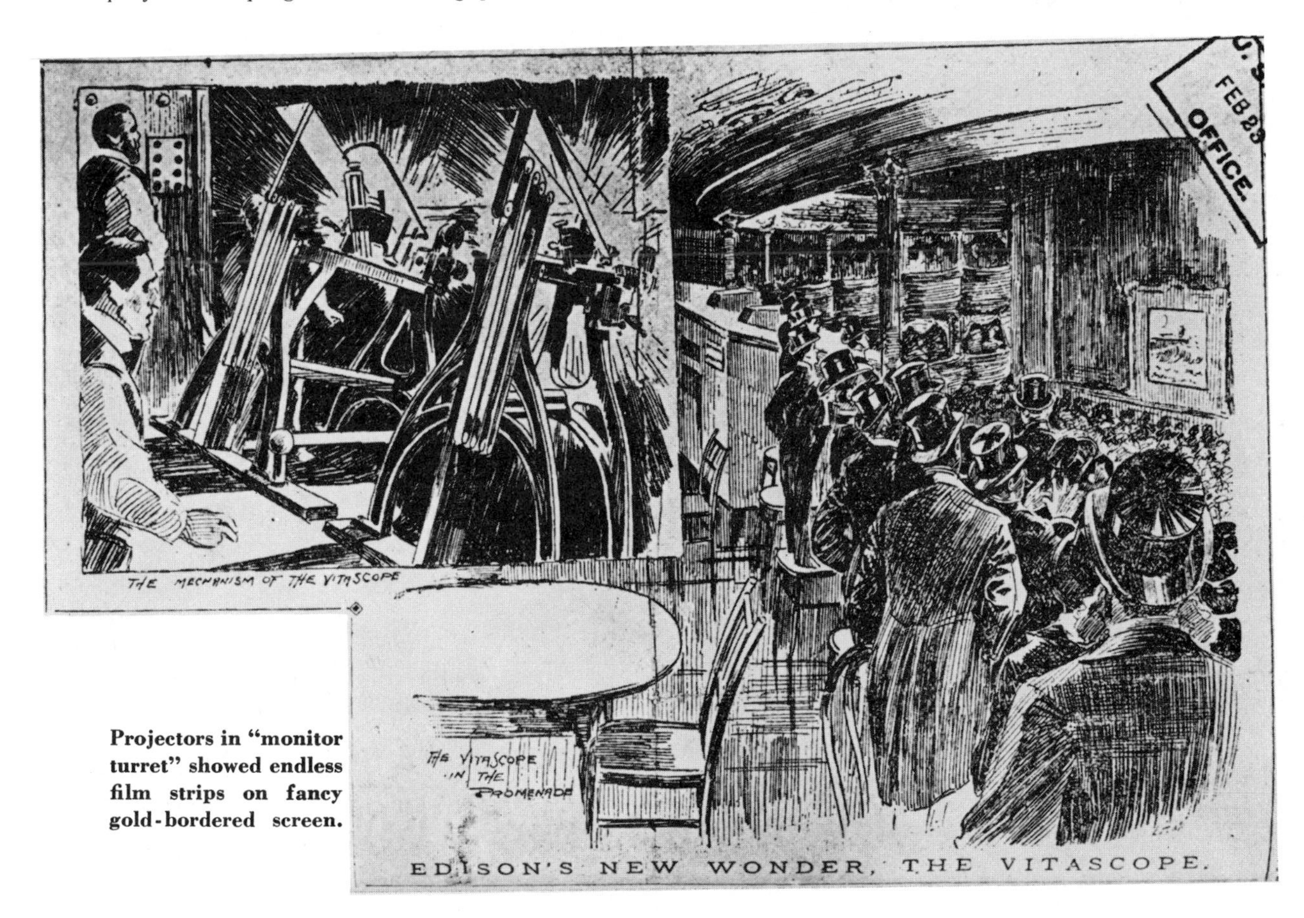

Projectors in "monitor turret" showed endless film strips on fancy gold-bordered screen.

Sunday night at the Bijou found Junior, Papa and Mama, Sister Sue and Chalk-stripe Charlie in the front row having one swell time...

in spite of threatened catastrophe...

Between reels a tenor in a nifty suit plugged the latest hits assisted by genuine hand-colored slides. In those pre-Bouncing Ball days, *he* did all the singing; patrons got the Pianola roll next day.

Some of the numbers were sentimental...

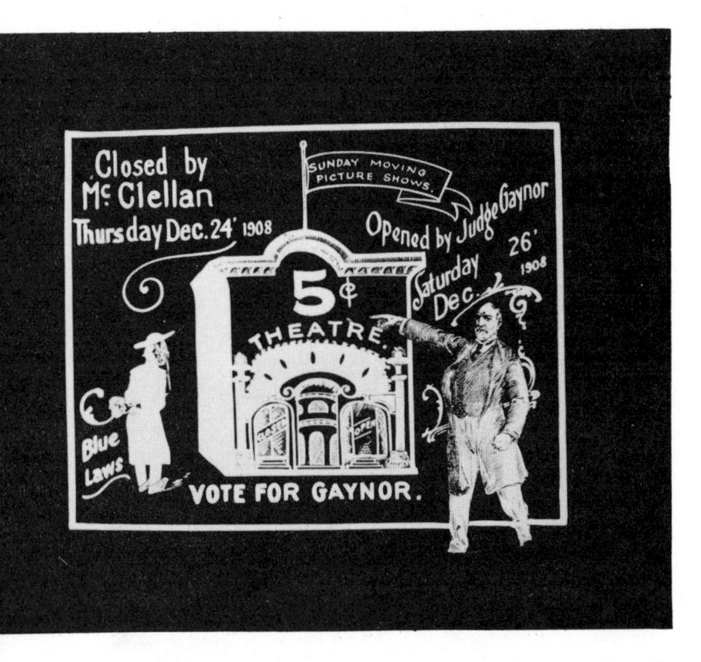

and the harassment of blue-nosed killjoys. (Hurrah for good old Gaynor, he saved the day!)

As business got better and prices doubled, fronts got fancier. The Unique, on New York's 14th St., sported a box office that *was* unique.

while others were pictures from life's other side...

but you could always count on a patriotic song before things settled down for the next reel, and this time the audience might be allowed to join in.

In New York everyone was movie mad. Marcus Loew's Herald Square Theatre (on the northwest corner of Broadway and Thirty-fifth Street) was reeling off daylight photoplays—the "silver sheet" was the secret—on a grinding thirteen-hour schedule. To add spice to his program, Loew hired "an out-of-work coon-shouter named Sophie Tucker" to sing fourteen times a day — seven shows in blackface, seven in whiteface. Sophie spent her time blacking up, coon-shouting, washing up, and coon-shouting some more as she alternated her way through the day's arduous chiaroscuro. The Herald Square (ex-Weber & Fields') was a former variety house that could seat more people at a single show (and at twice the price per head) than five ordinary nickelodeons. There was money to be made, more and more of it, and theatres, Mr. Loew—and Miss Tucker—began to get bigger and bigger.

Before long, real theatres were being built specially for movies. As the idea caught on, these became more and more lavish, with uniformed staffs, resident orchestras, squads of soloists, "mammoth" pipe organs, printed programs . . . and still the people clamored for more. More of everything. And they got it; by the end of World War I, they saw hundreds of handsome new (and profitable) theatres open their doors to booming business. The old nickelodeons were hanging on like aging courtesans, still grinding 'em out, still packing 'em in . . . for a nickel, but in 1919 there arose the first 5,000-seat colossus, a portent of things to come.

The trend was set, the time was right. The reel marked *The Twenties* was waiting to flash on the screen . . . and so was the golden age of the movie palace.

•

The Twenties were a time of extremes . . . extremes in wealth and poverty, culture and vulgarity, ambition and what-the-hell. The massive leveling processes of the Thirties had not yet begun to bulldoze away the social and economic differences that set people apart. The bastions of Society were still unscaled by the masses, and a

Featuring The Last of the Red Hot Mamas in black and white.

name like Vanderbilt meant a Fifth Avenue mansion, not an etiquette book.

It was an era of ferment and change . . . a wildly creative, slavishly imitative, manic-depressive, "I'll Build a Stairway to Paradise"—"I Faw Down and Go Boom" decade that has become—in the thirty years since it ended—the period best beloved by latter-day social historians and nostalgia merchants.

Life was not nearly so joyous for the *hoi polloi* in the Twenties as many of its chroniclers would have us believe. There were, of course, such fondly remembered institutions as Mah Jong, rumble seats, home brew, and doo-wack-a-doo. But for many people it was a time of creeping boredom and frustration. It was before the day of the bowling craze, the cook-out, the home workshop, and the lures of go-now-pay-later vacations. The Pianola and the Victrola were limited by the number of rolls and records in their cabinets; there was no television, and radio appealed only to the ear and was, at best, a stay-at-home attraction. Only one family in ten owned an automobile. And yet people were stirred—like lemmings—with an urge (to use a then-popular phrase) to "go places and do things."

And so they went to the movies.

Everywhere there was a thirsty curiosity about the lives of the rich and the surroundings those lives were lived in. Hollywood had already discovered this secret, and rags-to-riches was filmdom's bread and butter. As the movies created their own glamorous climate, smart exhibitors sought to capitalize on it. And theatres, with their peep-show days scarcely forgotten, were growing in sophistication along with their audiences. Films were Art . . . *Ars longa, vita brevis* . . . *Ars Gratia Artis*. And the theatres strove to keep pace with each new celluloid extravaganza.

Movie palaces sprang up in vast and shining numbers across the land; their towering electric signs—flashing "Paramount" "Paramount" "Paramount" "Paramount" or "Loew's" "Loew's" "Loew's" "Loew's" or "Fox"—were beacons in the darkness over a thousand Main Streets. They spelled out fun, enchantment, and escape to all the millions in the Twenties who wanted so desperately to believe in make-believe.

Here, at last, was Aladdin's arc-lit lamp. Here was the chocolate-covered lotus flower, the air-conditioned castle in Spain, the ageless Sirens accompanied by Mighty Wurlitzer. And all for twenty-five cents (before six o'clock).

So people went to the movies . . . religiously, once a week. Surrounded by forests of classic columns, armies of uniformed flunkies, galleries of oil paintings, and arcades of mirrors; seated in the violet twilight of huge French or Chinese or Italian or Moorish garden-auditoriums; soaking up music and laughter and mystery and romance in turn . . . here moviegoers found respite from reality, and a whole new dream world come true.

Here anyone with a little loose change might dwell in marble halls for a couple of hours. And the keener the competition among owners and builders, as more theatres arose, the more *marble* the halls became. Exhibitors vied for audiences with the energy of peacocks in a mating dance. Façades were set ablaze by marquees with colored electric lights racing around their borders. In summer, entire theatre fronts were frosted over like the ice compartments of neglected Frigidaires, to tout the arctic chill within. If the Rivoli hired a giant in a gendarme's uniform to guard the box office, the Tivoli promptly dressed a midget as a Keystone Kop to patrol the foyer, and the doorman at the Grand put on a burnoose and joined the Foreign Legion.

The people loved it. After all, it was for them that this sumptuous and magic world was built, and they thoroughly enjoyed being spoiled by indulgent impresarios. Ladies from cold-water flats could drop in at the movie palace after a tough day in the bargain basements and become queens to command. Budgets and bunions were forgotten as noses were powdered in *boîtes de poudre* worthy of the Pompadour. From a telephone booth disguised as a sedan chair, Mama could call home and say she'd be a little late and don't let the stew boil over.

October, 1929 . . .

A million lights they flicker there,
A million hearts beat quicker there.
No clouds of gray on the Great White Way –
That's the Broadway Melody.

CRITERION
THEATRE
MISS
HELEN MORGAN
IN
APPLAUSE
Paramount's
NEW SHOW WORLD
TALKING SENSATION
MISS
HELEN MORGAN
IN APPLAUSE
FRANCES WILLIAMS
TOM HOWARD
YACHT CLUB BOYS
Paramount's DRAMATIC
LOEW'S NEW YORK
PALACE
LOEW'S STATE

PROGRAMS FOR THE WEEK OF JUNE 11TH

With First Run Theatres

WIRE REPORTS FROM CORRESPONDENTS

NEW YORK CITY

The Rivoli Theatre—
Overture—"Sicilian Vespers" by Rivoli Orchestra.
Scenic—The Birth of an Iceberg.
Vocal—Aria from "The Masque Ball," sung by Marcel Salesco, baritone.
Current Events—Rivoli Pictorial.
Novelty—"Salterello," Italian dance by Amata Grassi and Senia Gluckoff, dancers.
Feature—Too Much Speed—Wallace Reid.
Special—Selections from "The Pink Lady" by Rivoli orchestra and Mary Fabian, soprano.
Comedy—Man vs. Woman—Christie.
Organ Solo—"Sixth Sonata" played by Prof. Firmin Swinnen.

Capitol Theatre—
Overture—"Capriccio Italien" by Capitol Grand Orchestra.
Scenic—"Venice"—Kineto.
Vocal—"Barcarolle" from "Tales of Hoffman" by Maria Samson and Erik Bye.
Current Events—Capitol News.
Musical—Salzedo Harp Ensemble, conducted by Carlos Salzedo, Marie Miller, First harpist, Elise Sorelle, second harpist, Edith Connor, Diana Hayes, Thurema Spear, Suzanne Bloch; (a) Celebrated Largo, (b) Spring Song; (c) Song of the Volga Boatmen.
Feature—Voice in the Dark—Goldwyn.
Special—Selections from "The Blue Paradise," by Capitol Grand Orchestra, Capitol Mixed Quartette, and Capitol Ballet Corps.
Comedy—Made in the Kitchen—Mack Sennett.

Mark Strand Theatre—
Overture—"Capriccio Italien," by Strand Symphony Orchestra.
Current Events—Mark Strand Topical Review.
Educational—"Outwitting the Timber Wolf"—Pathe.
Vocal—Strand Ladies Quartette, Irma Gallenhamp and Zellah Wilson, sopranos, Alma Keller, and Elinor Hughes, altos.
Feature—Lessons in Love—Constance Talmadge.
Musical—Violin Solo—"Concerto," by Katherine Stang, violinist.
Comedy—The Greenhorn—Educational.
Organ Solo—"Grand Choeur," by Fred. M. Smith and Herbert Sisson, organists.

Rialto Theatre—
Overture—"Wagneriana," by Rialto Orchestra.
Educational—Switzerland—Kineto Review.
Vocal—"The Want of You," sung by Robert White, tenor.
Current Events—Rialto Magazine.
Musical—"Broadway Hits" played on the marimbaphone by Harry Edison.

The California theatre, Los Angeles, billed "Salvage" as shown by this cut

Feature—One A Minute—Douglas MacLean.
Vocal—"Villanelle," sung by Grace Hoffman, soprano.
Comedy—Three Good Pals—Sunshine.
Organ Solo—"Festival Toccata," played by John Priest.

Criterion Theatre—
"White and Unmarried" in its second week.

SEATTLE

Coliseum Theatre—
Overture—The Fortune Teller.
Comedy—The Sneakers.
Current Events—Pathe News.
Special—Vocal number by trio
Feature—The Sky Pilot—1st National.
Next Week—Hush.
Clara Kimball Young appearing in person.

SAN FRANCISCO

California Theatre—
News—
Overture—Orchestra and Duo Art Piano—Tschaikowsky Concerto.
Digest Topics.
Organ—"Dance of the Hours" and "Wonderful Memories."
Special—McWilliams and Van in Harmony Numbers.
Feature—Charge It.
Next Week—Traveling Salesman and The Bakery.

Imperial Theatre—
2nd week of Deception.
Next—The Gilded Goose.

LOS ANGELES

Grauman's Theatre—
Overture—"Sweet Little Bon Bon."
News—Pathe.
Organ Solo—"Lonely."
Topics of the Day—
Vocal—Girl Singing "Bright Eyes."
Department Store Setting—Two Girls Original Composition.
Vocal—Tenor solo—"La Boheme."
Prologue—Garden Setting.
Feature—The Wild Goose.
Scenic — Educational—South Sea Magic.
Special—Specialty Dance and Mandolin Solo.

CHICAGO

Tivoli Theatre—
Overture—"Stradella."
Specialty—"Somewhere A Voice Is Calling."
Specialty—Holt and Rosedale sing-in "Hiawatha's Melody."
Feature—Bob Hampton of Placer.
Organ Solo—"Now I Lay Me Down To Sleep."

CINCINNATI

Capitol—
Overture—Oberon.
Capitol News Events.
Song presentation: Winds in the south.
To A Wild Rose.
Albert Bollinger, boy soprano.
Literary Digest—Topics of the Day.
Presentation: An Episode of the Race Track.
Ella Wheeler Wilcox's poem, "Salvator," reading by Dorothy Hecker; with orchestral interpretation and pictures.
Feature—Sacred and Profane Love.
Orgologue—Family Fantasies—Organ solo.
Comedy—P-A-L S—Century Comedy.

BALTIMORE

Century—
Overture—"Irene," by Tierney.
Current Events—Kinograms.
Added—Century Comedy News.
Special—Prizma.
Classic—Scene from "La Forza del Destino"; Justin Lawrie, tenor, and Fernando Guarneri, baritone, from Chicago and Metropolitan Opera Houses.
Feature—White and Unmarried.

WASHINGTON

Crandall's Metropolitan—
Overture—Tickle Me Selections.
Current News—Pathe News—Topics of the Day.
Comedy—The Skipper Has His Fling (Toonerville).
Feature—The Old Swimmin' Hole (First National).
Next Week—The Sky Pilot (First National).

Loew's Palace—
Overture—Sweethearts.
Current Events—Pathe News—Topics of the Day.
Comedy—The Jockey.
Feature—The Traveling Salesman (Paramount).

ATLANTA

Howard—
Overture—"Tales from Hoffman," Howard Concert Orchestra, conducted by Enrico Leide, and Vincent Kay.
Howard News and Views.
Violin Solo—"To A Wild Rose," by H. Hemery, concert violinist.
Novelty—Grantland Rice Sport Pictorial.
Comedy—Mutt & Jeff in "A Crazy Idea."
Feature—Thomas Meighan in "White and Unmarried."

CLEVELAND

Allen—
Overture—Fantasia Gypsy Life arranged by Musical Director Philip Spitalny, followed by a humoresque on "Margie."
Prelude—"The Street of Good and Evil" arranged by Managing Director S. Barrett McCormick and art director Frank Zimmerer.
Feature—"Dream Street"—United Artist.

ST. PAUL

State Theatre—

(a) Overture—" Masaniello."
(b) State Digest — A compendium of news events, educational and travel pictures and scenic pastels. Includes: Current Events —Kinograms and International.
(c) "New Wine in Old Bottles"—Chester Screenic.
(d) " Playmates "—Universal.
(e) Mabel Weeks and Bernard Ferguson:
  1. "Just That One Hour"—Eville—Bernard Ferguson, baritone.
  2. "Honeymoon Land" — Gillette—Mabel Weeks, soprano; Bernard Ferguson, baritone.
(f) Constance Talmadge in "Lessons in Love."
(g) Organ — " The Lost Chord "—Sullivan.
(h) Noon day Organ Recital, by Arthur Depew:
  1. Andante from "Ballet Egyptian"—Luigini.
  2. "Daybreak," from "Peer Gynt Suite—Grieg.

PHILADELPHIA

Stanley—

Prelude—" The Hunt is Up."
Special vocal number arranged by Hugo Riesenfeld and produced here by Josiah Zuro.
Characters:
A minstrel; the jester; couriers, and hunters. Singers, Carl Robbins, Ludwig Burgstaller and Rivoli Male chorus.
Feature—Deception.
Stanley News—Special compilation from all the weeklies of the world.
Scenic—British Castles.

KANSAS CITY

Newman Theatre—

Overture — " William Tell," by Newman concert orchestra, Leo F. Forbstein, director.
Newman News and Views.
Organ Selections—Gerald F. Baker and Quentin Landwehr, organists.
Special Number—Second Anniversary Frolic, with thirty-five people in prologue and special stage settings.
Feature—Too Much Speed, with Wallace Reid.
Next Week Lessons in Love.

ST. LOUIS

New Grand Central—

Overture—" If I Were King."—Rodemich.
Current Events—Selections from leading strips.
Musicale " Tittles Serenade." French horn and flute.
Feature—Lessons In Love—Constance Talmadge.
Organ solo—My hero—Yost at organ.

BROOKLYN

Mark Strand Theatre—

Marie Dvorak, pianist, daughter of Anton Dvorak, noted composer, in recital:
(a) Third Movement from Grieg's " Concerto."
(b) " Humoresque " (Dvorak).
Violin Solo and Dance.
Erminie Mathews, danseuse, in swan dance before rural drop with water ripple effect, with Jeno Sevely, violinist, furnishing accompaniment with Saint-Saens' " The Swan."
Concert Recital.
(a) " Siciliana " from Mascagni's " Cavalleria Rusticana," sung by Georges DuFranne, French tenor.
(b) " My Heart at thy Sweet Voice " from Saint-Saens' " Samson et Dalila," sung by Kathryn James, contralto.
(c) " Bird Song " from Leoncavallo's " Fagliacci," sung by Eldora Stanford, soprano.
(d) " Thou Sweetest Maiden," from Puccini's " La Boheme," duet sung by Miss Stanford and Mr. DuFranne.
Mark Strand Topical Review.
Prologue to Feature Film:
Artists singing before illuminated French doors of futuristic interior with transparent columns at front sides of stage.
(a) Mr. DuFranne singing " You're in Love," from Friml's " Loveland."
(b) Miss James singing " The Message of the Violet " from Luder's " The Prince of Pilsen."
(c) Eldora Stanford, soprano, singing " Sweethearts," from Herbert's " Sweethearts."
Feature—Constance Talmadge in Lessons in Love.
Overture — " Beautiful Galatea " — Suppe.
Special Film—A Day With Jack Dempsey.
Organ Solo—" Etude in G Flat Major "—Chopin.

One of the artistic hand drawn displays Harold B. Franklin, manager of Shea's Criterion theatre, Buffalo, used in billing " The Lost Romance "

By 1921 the movie palace idea had swept across the continent like, as the founder and former owner of Goldwyn Pictures put it, "wildflowers." Entertainment was blooming on a scale that the public had never seen before; symphony orchestras, dancers *da prim'ordine*, elaborate settings, big names, and carefully planned programs were a weekly occurrence wherever first-run movies were shown. The local opera house or vaudeville theatre could offer nothing as beguiling as the combination to be found at the movie palace on Main Street, as this round-up from the pages of *Motion Picture News*, on what was playing in a mid-summer's week, indicates.

ORIENTAL
GRILL
ALL SEATS 15¢

CAPITOL
PRETTY LADIES MLLE-GAMBARELLI
ORCHESTRA BALLET SOLOISTS
LARGEST THEATRE COOLING PLANT
CAPITOL
PRETTY LADIES
THE CAPITOL'S FAMOUS PROGRAM
COOLING PLANT

HOWARD
HAROLD LLOYD IN GIRL SHY LAUGH ROAR
C SHARPE MINOR WIZARD OF THE ORGAN

Across the nation, night and day, they packed 'em in. Bargain prices at the Oriental in Chicago, air cooling at the Capitol in New York, Dr. Robyn at the Grand Central in St. Louis, a world premiere at Grauman's Chinese in Hollywood, or a kiddie matinee at the Howard in Atlanta . . . whatever the attraction, people went out to the movies week after week.

For Mama, another world lay beyond the solid bronze box office where the marcelled blonde sat (beside the rose in the bud vase) and zipped out the tickets, sent the change rattling down the chute, read *Photoplay*, and buffed her nails—all without interrupting her telephone conversation. Heaven only knew what exotic promise waited behind the velvet ropes in the lobby, what ecstasy was to be tasted in the perfumed half-darkness of the loges. . . .

As she entered the Grand Foyer and surrendered her ticket to the generalissimo at the door, Mama could *feel*, more than hear, the rumbling majesty of the Mighty Wurlitzer in the still-distant auditorium. Not to be fooled by the dashing grenadier who chanted "For the best remaining seats, take the Grand Staircase to the left," she would tighten her grip on her shopping bag and forge ahead to the orchestra floor. Here she would be greeted by a cadet from the court of Franz Josef who would usher her into the auditorium with a deference usually reserved for within-the-ribbons guests at society weddings. Down the aisle he would escort her until just the right seat was found. Then, with a smile and a quick salute, the usher would vanish and leave Mama to settle back in the crimson plush, slip off her Enna-Jetticks and lose herself—body and soul—in Never-Never Land.

Many nights Mama came with the whole family. Usually a certain evening each week was set aside for going out to the movies, and usually they went to the same theatre every week. As the dishes were hurriedly dried and the milk bottles set out, nobody bothered to look at the paper to see what was playing. It really didn't matter; the picture on the screen at the Xanadu was secondary to the total adventure. It might be bad more often than not, but it was always over in little more than an hour, and then the fun began.

This time Papa bought the tickets and Junior scrunched down in his lumber jacket so he could get in for a dime. Mama and Sister, who had stopped by Liggett's to buy Necco Wafers and Tootsie Rolls (this was before the days of candy counters, and popcorn was something you got only at the circus), joined them in the outer lobby. For Junior (and secretly for the others) this stroll along the polished terrazzo, across the colored rubber mats, past the bronze showcases with the stills for next week's show mounted behind cutouts in the silver-flittered compo boards, past the fountain where giant goldfish who had never seen the light of day swam slowly beneath the changing colored spotlight, past the polished brass doors, past the ticket-taker beside his ornate gold ticket chopper, and finally into the Grand Foyer . . . this was a journey of almost insupportable suspense. Then, if the orchestra could be heard playing mutedly beyond the glass partitions at the back of the auditorium (the stage show had already begun!), Junior could stand it no longer. He broke ranks and ran for the Grand Staircase and the balcony that lay beyond.

In the evenings they always sat in the balcony. It was so nice to be able to look up and see the stars twinkle, even if it was raining outside. And those lovely clouds . . . how *did* they do it? Mama never knew.

So there they were. Who cared if the ferns in the porch box at home were dying or if the Essex needed new tires? They were sitting in the midst of a crowd of happy people in the most beautiful place they had ever seen, while way down below a stageful of talented performers was going to see to it that they had a wonderful time.

First came the congregational singing of June-moon-soon songs led by the bouncing ball and accompanied by Melody Mac at the Mighty Wurlitzer. Then the Xanadu Grand Orchestra would rise majestically on its platform in an effulgence of harps and horns and "Reminiscences of Strauss." Next the beautiful curtains with the rhinestone butterflies would open on the stage presentation with its girls on giant staircases (or girls on unicycles, or girls playing fifteen white baby grands), followed by tuxedoed tenors, wonder dogs, tap-dancing moppets, jazz bands in funny hats, and adagio teams. Then the short subjects with their grab-bag of newsreel, comedy, travelogue and educational novelty. Even the "coming attractions"

held promise of fresh delights only a week away. And if the picture turned out to be a mushy old love story, Junior didn't mind; he could always snap Necco Wafers and watch the dull blue sparks flash in the dark . . . or go to the Gentlemen's Smoking Lounge and run the electric hand dryers . . . or count the stars . . . or figure out about those clouds.

THE DREAM BEGINS

S. L. ROTHAPFEL

"A child of emotion, living in a self-centered world bedecked with iridescent dream-colors, where impersonal fact was as remote as the fifth dimension, he never knew from whom or where or why came the blaze of the lights of fame — nor why they dimmed."

TERRY RAMSAYE, in a tribute to Roxy, in *Motion Picture Herald*, January 18, 1936

In 1961 the telephone books of ten of the largest cities in the U.S. listed 144 businesses of all descriptions under the name "Roxy." Delicatessens, dress shops, dry cleaners, hotels, bowling alleys, printers, butchers, bakers, and button makers—and theatres—all named after a man who had been dead twenty-five years, a man few of the proprietors had ever known or seen or even heard of.

Roxy . . . if ever there was magic in a name, this one had it.

•

Samuel Lionel Rothapfel (the "p" was dropped years later when Germanic names went out of fashion) was the second son of an immigrant German shoemaker and a Polish mother who had settled in the little lumber town of Stillwater, Minnesota, not far from St. Paul. There Sam was born to Gustave and Cecelia Rothapfel on July 9, 1882, and spent the first twelve years of his life doing all the things a frisky small-town boy did in those days—sliding down Morris Street in the winter, swimming and fishing in the St. Croix River in summer. But suddenly Sam and his brother Max found themselves in quite a different environment; Gustave Rothapfel decided in 1895 that the sidewalks of New York offered more opportunity than the sidewalks of Stillwater, which, if not actually rolled up at night, were certainly easier on shoe leather. So the family moved to Manhattan and settled, along with so many others like them, in the roiling, rootless slums of the Lower East Side.

Once in New York, Gustave sternly demanded that Sam put away childish things and make something of himself; his vision of life for the boy was a reflection of his own, and he tried to force him into the picture as he saw it. But Sam was dimly conscious of things outside that picture . . . and it was enough to keep him in constant rebellion. Sam did try. He got a job as a two-dollar-a-week cash boy in John B. Collins' department store on Fourteenth Street, a job that lasted four dollars' worth. After that it was a long series of odd and ill-paying jobs that he managed to hold for a week or so before his daydreams got the best of him.

Sam's mother always defended him—though she didn't understand him. It was the almost classic story of the inflexible patriarch, the loving but ineffectual mother, the ne'er-do-well dreamer of a son . . . a story that Samson Raphaelson immortalized years later in his play, *The Jazz Singer*. Mama Rothapfel died when Sam was sixteen, but Sam had really lost her two years earlier, when Gustave, his patience at an end, "chased him out of the house" . . . the black sheep driven from the fold.

More odd jobs . . . messenger boy, bootblack, callboy in the theatres and music halls that lined Broadway from Fourteenth Street to Herald Square. (Maybe Sam Rothapfel was on the scene that memorable night at Koster & Bials'.) Then at eighteen, alone and cut loose from any real purpose, he enlisted in the Marines and discovered, for the first time, the meaning of organized discipline — sternness with a reason he could grasp. He loved the Marines, and for seven years, including some real action in China during the Boxer Rebellion, he worked earnestly to become the best Marine in the world. He ended up—proud of his uniform, proud of the Corps—having a glorious time yelling his head off as a drill sergeant in the Dry Tortugas.

When his enlistment was up, he knew that he wanted something else . . . but he still didn't know what it was. He found himself starting to drift all over again. First he was a hash-slinger in a cheap hotel, next a clerk in a shoe store, then a hack driver for a livery stable. Somehow he wandered down into the coal-mining area of Pennsylvania, and tried out for a place on a struggling baseball team in the Northeast Pennsylvania League. This was another turning point, and here he got something that was to be one of his most valuable assets for the rest of his life: his magic nickname. When he would knock one over the fence, you couldn't expect the bleachers to yell, "Slide, Rothapfel, slide!" So "Roxy" it became, and Roxy it was to remain (though it was only a private sort of nickname until years later, when the whole nation took it up).

In the fall of 1907 the Northeast Pennsylvania League went into winter quarters (the local coal mines), and Roxy turned to door-to-door peddling. His wares were a set of illustrated travel books—*Stoddard's Lectures*—available in a variety of ornamental bindings, none of which appealed to coal miners.

One December day he trudged discouragedly into Forest City, Pennsylvania, a little mining town on the banks of the Lackawanna River. Stopping at the town saloon, he tried halfheartedly to unload a set of *Stoddards* on the proprietor. Hardened, by this time, to failure, he decided to celebrate his sixth "no sale" for the day with a mug or two of beer and some frankfurters, of which he was inordinately fond. Through the open door of the kitchen he spied the publican's comely daughter, Rosa. The longer he stayed munching the excellent hot dogs, drinking the good beer, soaking up the warmth of the big cannon-ball stove, and stealing occasional glances toward the kitchen, the better Roxy liked Forest City. Why not dump *Stoddard* and his *Lectures* for good and apply for a job behind the bar?

Rosa's unsuspecting father, Julius Freedman, happy to have such a literate and well-traveled gent to ornament his establishment, agreed. Julius, who was Forest City's banker, innkeeper, and undertaker, was a one-man civic-improvement committee, and any new arrival who was not a coal miner was more than welcome. That afternoon, in white apron and sleeve garters, S. L. Rothapfel found himself selling something that people *wanted* to buy. This had never happened to

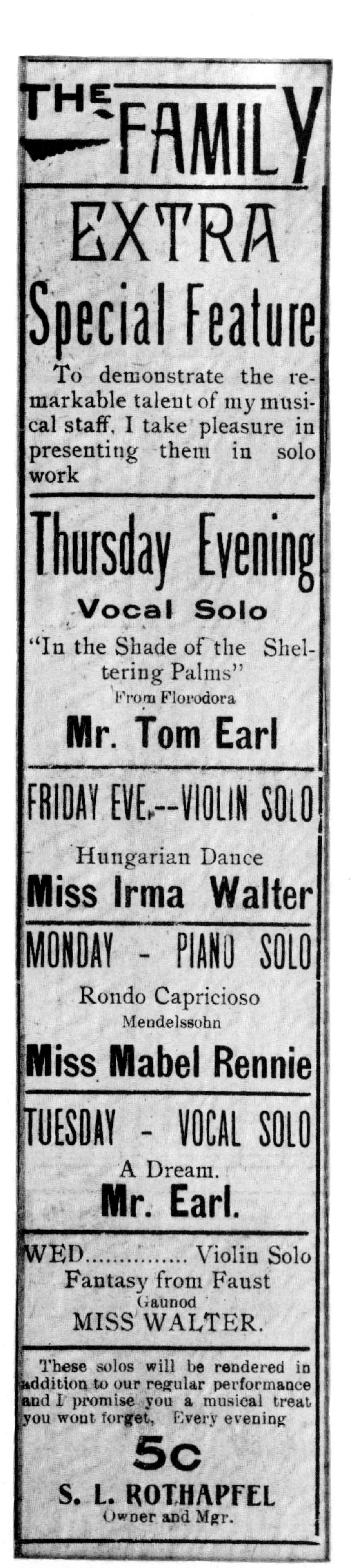
THE FAMILY

EXTRA

Special Feature

To demonstrate the remarkable talent of my musical staff, I take pleasure in presenting them in solo work

Thursday Evening

Vocal Solo

"In the Shade of the Sheltering Palms"
From Florodora

Mr. Tom Earl

FRIDAY EVE.--VIOLIN SOLO

Hungarian Dance

Miss Irma Walter

MONDAY - PIANO SOLO

Rondo Capricioso
Mendelssohn

Miss Mabel Rennie

TUESDAY - VOCAL SOLO

A Dream.

Mr. Earl.

WED............... Violin Solo
Fantasy from Faust
Gaunod
MISS WALTER.

These solos will be rendered in addition to our regular performance and I promise you a musical treat you wont forget, Every evening

5c

S. L. ROTHAPFEL
Owner and Mgr.

him before; it was a sensation he liked very much indeed.

Behind Julius Freedman's tavern was a large room used sometimes for dances and sauerkraut suppers when the miners got their pay. Most of the time it just gathered dust. The idea struck Roxy that this might be a fine place to show the new "living pictures" that had become such a fad in the larger towns. Even McKeesport and Newcastle had them; why not Forest City?

Rothapfel's Family Theatre—named, perhaps, because its door was already marked "Family Entrance"—opened with roller skating, not movies, for some reason. A week later, on Christmas Eve, 1908, Roxy got his booking problems untangled and ran an ad in the Forest City *News* announcing "Grand Opening in Vaudeville—Special Matinee Xmas Afternoon." Rosa's papa loaned the chairs from his funeral parlor . . . subject to sudden recall; the screen was a bedsheet, and the projector was a battered, hand-cranked Cineograph. On opening day, Roxy was at the depot in Carbondale at four-thirty a.m. to meet the milk train bringing his first reels of film. The first program consisted of an overture by Prof. James Curry, pianist, and a mixed bag of vaudeville, illustrated songs, a one-reel comedy, and the feature—films of "the World's Series between the Chicago 'Cubs' and the Detroit 'Tigers,' showing the famous plays and all the favorites in actual motion." The Family was billed as the "coziest and best equipped in the Valley," but reserved "the right to eject any undesirable or disorderly characters from the premises."

The Family's staff consisted only of Roxy and the devoted Rosa, who had since become his bride. Roxy singlehandedly painted the posters, distributed the handbills, picked up the films, ran the projector, and booked the "talent." But though its staff was small, the Family Theatre had class.

An early Family Theatre handbill illustrates rich variety of musical life in Forest City.

Roxy was constantly experimenting and dreaming up new ideas to please the patrons: first a curtain that closed over the screen between reels, then a system of colored light bulbs concealed around the screen that changed from pink to green to blue as Roxy threw the switches during the film. Miss Rennie was augmented by a violinist and a baritone who offered classical selections while the films were being changed.

Along about this time New York film entrepreneurs began to notice that an exhibitor in the sticks somewhere way down in Pennsylvania was renting as many reels as the busiest nickelodeon operators in the city. And word began to filter back up along the line of the film exchanges that something pretty special was going on in the Family Theatre in Forest City. Before long certain city types were arriving to see what was what, and word of the "high-toned little theatre" reached the redoubtable Benjamin Franklin Keith, grand vizier of U.S. vaudeville.

Keith, who didn't think much of movies himself, was running off a reel or two of cheap film at the close of his variety bills in an effort to drive patrons from the theatre and clear the house for the next performance. But his conscience hurt him and he had come to the conclusion that, if he were going to show movies at all, he might as well do it right. So he sent for Roxy and commissioned him to improve the motion picture presentations on all Keith vaudeville bills. Roxy did such a good job that soon people were staying just to see the movies, and Keith had reason to ponder the creation of this flickering Frankenstein monster.

During the Keith period Roxy traveled all over the country. One day, on the train coming down from Minneapolis (where he had just been helping the manager of the Lyric Theatre glamorize the movie portion of his vaudeville bill) Roxy happened to meet a gentleman named Herman Fehr of Milwaukee, and they struck up a conversation in the club car over their mutual interest, theatres. Fehr owned a large and handsome house in Milwaukee called the Alhambra—large, handsome, and a flop as a money-maker. After several hours' conversation during which Roxy outlined some of his ideas on music and lighting and theatre operation, Fehr persuaded him to come to Milwaukee and see what he could do to ensure that if the Alhambra *must* lay eggs they would be golden.

Roxy was on his way to Chicago, where another Lyric (this one a Shubert house) needed his help. After a short but effective ministry in the Windy City, he headed back to Milwaukee to bring salvation to the wayward Alhambra. Fehr met him at the station, and together they went to the theatre and sat in the empty auditorium while Fehr repeated the story of its failure as a playhouse. He tried to book the finest attractions, he said, but the competition in Milwaukee was just too keen. Finally Roxy turned to him:

"The Alhambra can be made to pay . . . but in just one way."

"How?" demanded Fehr.

"Motion pictures," Roxy said. "Movies."

Fehr was aghast at the idea of sullying his Alhambra with anything so raffish as movies, but Roxy convinced him that he meant *photoplays* (the first feature films, telling complete stories—often quite creditably—were just beginning to appear), not nickelodeon fare, and before they parted that afternoon Fehr had given Roxy a check for five thousand dollars and told him to make any changes he wanted to the Alhambra . . .

just as long as he didn't close the house to do it. Fehr had actors under contract and couldn't afford to close until the contracts had run out.

So Roxy began his work in the theatre's off hours. First he installed a nursery at the rear of the auditorium, where matineeing Milwaukee mothers might park their children while they enjoyed the movies. The rest of the theatre was refurbished with amber electric lights, new plush draperies, and carpeting on the floors. A staff of ushers was hired, fitted with handsome uniforms, drilled in courtly manners. A projection booth was built in the balcony. The orchestra pit was covered over; Roxy explained that the Alhambra orchestra was going to be installed on the stage.

"For all Milwaukee to see?" asked Herman Fehr. "Up there, sawing away?"

"They're expensive musicians," said Roxy.

And so the Alhambra orchestra was seated in a sort of Japanese garden, surrounded by potted palms and artistic, hand-painted scenery. After the special lighting effects were put in, even Fehr agreed that the results were spectacular.

Someone else had been watching all these changes with a practiced and appreciative eye. Mme. Sarah Bernhardt was appearing at the Alhambra on one of her annual farewell tours, and one morning the Divine Sarah summoned Roxy to her dressing room. Through an interpreter she had him explain what he had been doing, what his ideas and plans were. Her knowledge of the techniques of stagecraft amazed Roxy, and he found himself pouring out all his dreams of music and rainbow colors and clouds of gold in terms of descriptive scores, spotlights and gelatins, satins and scrim. When the interview was over, Mme. Bernhardt allowed him to kiss her hand (Roxy was learning fast) and said: "Vous irez loin!"

•

"You've gone too far, this time!" shouted Herman Fehr when Roxy outlined his final coup before the Alhambra reopened. "You've bank rupted me." But the unperturbed Roxy sent out a letter to a thousand members of Milwaukee's beer peerage, and other citizens of influence. It read:

Dear Sir:

Enclosed please find two tickets for the New Alhambra Theatre, which will admit yourself and a friend to see the beautiful new theatre and the wonderful Italian production, *The Fall of Troy*; which cost the producers more than $30,000.

Please do not feel that you are in any way obligated to this theatre. We merely wish to show you what remarkable advance has been made in motion photography and what a valuable adjunct the moving picture has become as an educator.

Respectfully yours,
S. L. Rothapfel

Fehr needn't have worried. With Rothapfel's elegantly composed letters, Rothapfel's choice of a feature film produced at such an astronomical cost, and Rothapfel's head full of big ideas, the Alhambra began paying a handsome profit the first week. Three months later the Saxe Brothers, operators of a flourishing chain of Milwaukee movie houses, paid Fehr $40,000 just to get control of the Alhambra.

But Roxy wasn't satisfied. His destiny, he knew, lay in New York. He had heard of Marcus Loew and his fast-growing chain of picture houses, and decided that he must go to New York and have a chat with him. Perhaps Loew, too, had an ailing theatre that Dr. Rothapfel could cure. So, in the autumn of 1913 he started out on almost the same journey he had made eighteen years before, when he was a kid, traveling with his family to Manhattan's promised land. He grew more and more excited as the train clipped along.

"This time I'm not going to be anybody's two-dollar cash boy," he told Rosa, who faced him on the green plush Pullman seat beside the two children, Arthur, four, and the baby, Beta. (Had Arthur escaped being Alpha by being a boy?) "I'm going to be the manager of a theatre. I don't know what the name of it is — but it's going to be famous, and so am I."

•

The famous theatre was the Regent, on the corner of 116th Street and Seventh Avenue. In 1913 the Regent had a real glory; it was the first de luxe theatre built expressly for showing movies in New York, even though Thomas W. Lamb had designed it to differ little physically from the many other pretentious neighborhood vaudeville houses that had sprung up around the city in the heyday of two-a-day. But even before it was built, its owner—Henry N. Marvin, scientist, college professor, and founder of the prosperous Biograph Company — had made it plain that it was to be dedicated to the *movies*, not vaudeville. And no wonder; with B. F. Keith's popular Alhambra Theatre right in the neighborhood, the Regent had to be different to survive.

Mr. Marvin somehow guessed wrong. Neighborhood vaudeville was a huge success all over town, with the biggest headliners of the day making the circuits from one popular theatre to the next. The Colonial, the Riverside, the Hamilton, the Audubon, the Orpheum—all of them miles from Times Square—were doing handsomely. But neighborhood movies were another matter. Most of them were throwbacks to the nickelodeons of earlier days, and while they were profitable, audiences were made up mostly of kids and people who wanted to kill an hour with a nickel . . . and nobody took them very seriously.

The Regent took itself *very* seriously. It had opened in February 1913 with an eight-piece orchestra, a separate string ensemble called the Mendelssohn Trio (violin, piano, bass viol), and Konrad C. Korschat at New York's first movie pipe organ. The opening program was made up of popular musical selections and features in Kinemacolor, a process that projected black-and-white films through colored filters to achieve a picture-postcard effect often amazingly true to life. Kinemacolor was a startling innovation in 1913, and when projected on another innovation, the Regent's "made-to-order picture curtain" (screen) with black borders that made the picture seem brighter and better defined, the effect brought loud cheers from the critics.

But it was all too much for 116th Street. Harlem, in those less complicated days, was still the seat of a stolid first-generation German *Burgherkind*. They were impressed by the Regent's opera-house splendor, with its gilded stage boxes and fancy curtains, but they were embarrassed and baffled to find all this grandeur surrounding something as unimportant as movies. The rollicking vaudeville shows at Keith's Alhambra, on the other hand, were something they could cope with on their own terms, and after everybody had come in once for a look, business at the Regent started to drop off.

It wasn't because Claude Tally, the Regent's manager, didn't try. He gave them printed programs, had them shown to their seats by well-mannered ushers, and gave organist Korschat and the Mendelssohn Trio billing on the posters outside. After Kinemacolor failed to capture his audiences' fancy, he tried programs of first-run black-and-white films . . . one bill a few weeks after the Regent opened included Kalem's *The Siege of St. Petersburg* as the feature, a Vitagraph John Bunny comedy, a Pathé jungle scenic, an Edison comedy (*After the Welsh Rarebit*) and a Biograph featurette, *My Hero*. He tried balladists and popular singers—even "coon shouters," but in vain.

In April he put a stock company on the stage to bolster the film program. Their first offering, *For Revenue Only*, was a melodrama which caused *Variety's* critic to remark that "a four-actor stock outfit doesn't seem to fit into that swell movie." The neighborhood seemed to agree, and in June "that swell movie" decided to forgo movies altogether when it was announced that the Regent was negotiating with the Shuberts to stage musical comedies in the theatre with a change of bill weekly.

Things got worse at the Regent, but its owners were determined to stick it out for the rest of the year; too much money was tied up in the expensive house to think of throwing in the towel. The downtown critics continued in their acclaim of the theatre and its operation, and H. N. Marvin wished that he knew some way to pick the whole thing up and roll it seventy blocks nearer Times Square, where the public appreciated the finer things. Those dunderheads in lower Harlem didn't know a good thing when they had it right on their well-scrubbed front stoops.

•

In October of 1913, Roxy had scarcely heard of the Regent, let alone its troubles. He had come to New York to see Marcus Loew, and he had his eye on a Loew theatre in the heart of the city which he could use as a showcase for his new brand of Rothapfellian magic.

An appointment was made, and at the proper hour Roxy appeared in the Loew offices in the old Putnam Building on Times Square. Holding his straw hat between sweating hands, he sat on the hard oak bench in the outer office and looked down on Broadway while waiting to be ushered in.

From behind the frosted-glass door marked "Mr. Loew" came the sound of a fist being pounded on a desk, voices being raised in anger. The stenographer blushed into her Oliver upright; the office boy winked knowingly at the apprehensive Roxy. As the recriminations and the poundings grew louder behind the door, Roxy made a decision. Jamming his hat on his head, he bolted from the Loew office and headed for the comparative peace and quiet of Broadway as fast as his legs and the bird-cage Otis would take him.

On the street he bumped into Henry N. Marvin, whom he had met the day before on a courtesy call at the Biograph offices. It was a case of you're-just-the-man-I-wanted-to-see; Marvin had been investigating Roxy's exploits in Milwaukee and had decided that if anybody could bring the tottering Regent back to health, S. L. Rothapfel could.

On a hack ride through Central Park, Marvin outlined his problem and his proposition. By the time they emerged at 110th Street, Roxy had begun to realize that the "famous theatre" he had come to New York to manage might be only six blocks away.

When he got to 116th Street, Roxy sized up the neighborhood in one afternoon; the German-Americans who lived there were a lot like his own parents, and a lot like himself. Somewhere, floating in the sour cream and dill sauce of their *Gemütlichkeit* was a yearning for real *Kultur*. And *Kultur*, decided Roxy, who had barely squeaked through grammar school, was what they were going to get (and like) at the Regent.

First Roxy closed the theatre and set about a program of extensive—and expensive—alterations. He had the projection booth moved from its perch at the summit of the balcony to a position at the rear of the orchestra floor. This permitted almost ideal pictures by doing away with that bugaboo of all high-altitude projectionists: "keystoning"—the distortion (and loss of reflective light) of the picture as it hits the screen at an acute angle from above. The old projection booth was converted into an office for the new director of the Regent, and a private elevator was installed for his convenience.

On the drafty stage he built a permanent setting for the screen, described this way in a contemporary issue of *Motion Picture News:* "The stage of the Regent represents a conservatory. The musicians play on the stage and are partly hidden by potted plants and other decorations including an electric fountain. Two large windows open on either side of the screen, with appropriate backgrounds, giving a perspective to the scene. Here the changing lights of varied colors procure a most wonderful effect. A heavy velvet curtain is lowered in front of the screen between the pictures, when the singers appear at the windows or the orchestra plays a special selection. It is art in every sense of the word, and there is no wonder that a manager paying so much attention to all the details is rewarded by constantly increasing patronage."

Rothapfel doubled the size of the Regent's orchestra and began assembling a music library of opera-house pretension. His theory was the revolutionary one that the musical accompaniment for a film should fit the action on the screen, instead of the totally unrelated *Poet and Peasant* sort of thing that had then been passing for movie music, whether the film be a comedy, a drama, or a scenic subject. Before Roxy allowed a film to be shown, he painstakingly scored it with an accompaniment to fit the action and mood of each scene, drawing from Beethoven or Herbert and leaning

A MOVIE IN THE RIGHT DIRECTION

"Come on in, Mamie. The pictures ain't much good, but it sure is the best Perfumed theater in the city."

most heavily on the Russians. There was grumbling among the musicians, most of whom were vaudeville pit men. What little movie work they had done usually consisted of playing gems from *The Bohemian Girl* with all the repeats, straight through, and then back to the top again. This they could do from memory while reading *Captain Billy's Whiz-Bang*. But Roxy had a way with musicians, and when the theatre reopened, the Regent Symphony was as serious a group as any in town.

The Regent, under Roxy's guidance, was an instant success. He was careful to see that curious "downtowners" didn't crowd out the equally curious 116th Street *Familiengruppe* on opening night. The feature film was an important one: George Klein's production of *The Last Days of Pompeii*. David Belasco, who was present, paid close attention to every detail of Rothapfel's presentation. At the end of the evening he declared: "It is one of the best things I have ever seen."

The local members of the audience clucked their tongues over the stage fountain "with real water running up from out." They were dazzled by the brilliance of the screen, particularly in view of the fact that the auditorium was lit by rose-tinted lights all during the picture. Roxy was one of the first exhibitors to abide by the new Moving Picture Theatre Code that required theatres to be fully illuminated at all times. Nearly every other picture show in town was as dark as a whale's belly. And as redolent. The Regent's ventilating system was one of Roxy's prides, and certainly there was no need in his theatre for those functionaries familiar to nickelodeon audiences who went up and down the aisles squirting noisome heliotrope-*avec*-creosote into the air from spray guns.

Motion Picture News for December 6, 1913, covered the opening under the headline, "A DeLuxe Presentation."

A remarkable incident in the history of the motion picture took place Monday evening of last week at the Regent Theatre, 116th Street and Seventh Avenue, New York City. This sounds serious, and the writer so intends it.

The picture was *The Last Days of Pompeii.* This excellent production Mr. Rothapfel centered in an environment so pleasing, so perfect in artistic detail, that it seemed as if the setting were a prerequisite to the picture, that to an educated audience the two should, and must hereafter, go together.

Mr. Rothapfel also achieves a theme in his music. There is the same unity throughout the admirable score with which the picture is accompanied. But predominating, woven delicately here and there in the score, is the soft and beautiful song from *Aida* symbolic of the tender love story of the picture. The curtain rises to an inspiring prelude from the Regent pipe organ, rendered by Mr. Drew, an accomplished organist. Then, with a flare from the orchestra, the dark red velvet curtains before the screen are parted to admit the figure of an actor arrayed in Grecian robes. With excellent intonation he announces the opening scene of the picture. Later, just before the

The Regent in 1913 was New York's (and Roxy's) first de luxe house.

Roxy moved the projection booth to orchestra floor...

dressed his projectionists

Vesuvius scene, he appears again and for an interlude recites the thrilling lines from Bulwer-Lytton's novel which tell of the mob and its frenzied attack upon Arbaces.

At other intervals the monotony of the "silent stage"—there is bound to be some monotony in the long picture, however inspiring it is—was delightfully broken by the voices of trained singers from the windowed recesses above and at each side of the stage. Soft lights were played upon these windows and also upon the fountain which plays just in front of the orchestra platform.

The frenzied scenes in the doomed city following the eruption of the volcano were made most realistic by the accompaniment from *Lohengrin* and by a chorus of shrill voices back of the screen.

Mechanically as well as artistically, the presentation was flawless throughout. It was, from every standpoint, the best that has been seen in this city.

The handsome and comfortable interior of the Regent has much to do with the success of the performance. There is no finer theatre in New York in point of construction, and Mr. Rothapfel's skillful attention to details has given the interior a refinement not to be equalled in a single other theatre here.

The audience, it should be noted, while made up of persons living in the neighborhood of the Regent, was of the kind to be found in the best playhouses. Judged by their decorum and sincere appreciation, they might have been at the opera.

There is a single criticism to be entertained, and that has nothing to do with the performance itself. This concerns the price of admission. It should be twenty-five instead of fifteen cents—that is certain. And perhaps one might go farther and say that such a production should be on Broadway at much higher prices.

Mr. Rothapfel's answer is: "It will be."

And it was.

like brain surgeons...

and proved that musicians should be seen as well as heard.

The Strand hit Broadway like a bolt of lightning (thunder courtesy of S. L. Rothapfel).

"A WONDERFUL AUDIENCE IN COSTLY TOGS"

"I don't like to be following in everyone's footsteps. I don't like to do things the way everyone else does them. All you hear about these days is the everlasting cry of theatre managers that they are looking for 'what the people want.' That idea is fundamentally and disastrously wrong. The people themselves don't know what they want. They want to be entertained, that's all. Don't 'give the people what they want' — give 'em something better."

S. L. ROTHAPFEL
in an interview in *Green Book* magazine, 1914

In less than six months after his triumph uptown, Roxy was on Broadway, true to his word. But the events leading to his arrival there were somewhat complex.

The Mark brothers, Mitchell and Moe — late of Buffalo and a lucrative nickelodeon business — had leased the site of the old Brewster carriage factory on the corner of Broadway and Forty-seventh Street, to erect a giant "million-dollar" theatre. Their original plan, as announced in the papers, had been to build a fifty-cent vaudeville house, seating 2,800. But they lost interest in this idea when William Morris beat them to it with his Wonderland Theatre at Broadway and Forty-fifth Street — a Longacre Square Luna Park that combined bazaar and indoor fair, amusement rides for children on the roof, with continuous movies and music-hall acts in the theatre proper. Three months after announcing their first plans (January 1913) the Marks revealed that they were thinking now of making their new theatre a sort of Hippodrome operation (with a water tank on the stage and menagerie in the basement) where shows would dazzle the Broadway public at ten cents and twenty cents a head. But nothing else came of this plan.

Meanwhile, the Broadway public was beginning to wonder what — if anything — was happening on the corner of Forty-seventh Street. Since the demolition of the carriage factory, a large hole had appeared in the ground, but now no activity at all could be seen behind the hoarding that surrounded the property. By midsummer, knothole peepers reported that foundations for *something* had appeared, and in July, Mitchell and Moe revealed to the press that the new theatre was to be dedicated as "a temple of Thespis where New York's masses might witness the finest dramas from all ages, at prices no higher than the meanest music hall might ask."

In October four walls loomed above the sidewalks and another change in policy was announced — this one something of an eye-opener: the Metropolitan Opera, no less, they told reporters, was going to forsake its headquarters "down on 38th Street, an area fast going out of fashion" and move rig, wig and swan boat to the Mark brothers' new theatre. The previously announced seating capacity would, of course, be doubled in order to accommodate opera audiences, but the problem of boxes (or rather the lack of them) didn't come up.

This time the brothers were really talking out of the tops of their hats. Thirty-eighth Street was far from going out of fashion; in fact the area between Herald Square at Thirty-fourth Street and Times Square at Forty-second was considered New York's "Rialto" and was lined with busy and successful theatres. Forty-seventh Street — Longacre Square — where the Marks were building their whatever-it-was, was considered a new frontier; it was doubtful if anyone would journey that far up Broadway to see even a "Tom" show . . . and certainly not grand opera. When he heard the news, Signor Gatti-Casazza, the Met's volatile impresario, fired a volley of denials from Thirty-eighth Street that made the unfinished, cream-colored brick walls at Forty-seventh Street tremble on their foundations. "The Metropolitan Opera move uptown? *Non mai . . . giammai . . . buffoni insolenti!"*

In December, with the theatre nearly finished, Mitchell and Moe still didn't seem to know what to do with it. A deal with the San Francisco Gaiety Company to stage musical comedy and operetta in the house had just exploded in their

"At the Sign of the Lipstick!"

—an innovation in motion picture theatres has just been completed and opened to the ladies, patrons of the Mark Strand Theatre. It is called the cosmetics suite, and the lounging room is of satinwood and rosewood, with gold leaf on hand-carved decorations, and furnished in Louis XVI furniture and tapestries. The cosmetics suite, like all other features originated by the Mark Strand, is being copied by motion picture palaces. Mr. Richard Barthelmess and Miss Marion Coakley, playing in "The Enchanted Cottage," presided at a formal tea in opening the suite to the public.

At the Strand, Roxy put the orchestra on the stage again, with balconies and niches in the romantic Spanish set reserved for soloists. The screen was tucked away somewhere beyond those gates. Milady's cosmetic suite (below) was fit for a movie queen.

faces. Perhaps, hinted the press, motion pictures would be presented there when the theatre opened in the spring.

Perhaps. It seemed a bit audacious. But Mitchell and Moe Mark, in spite of Gatti's condemnation, were nobody's *buffoni.* Movies, of course, were not new to Broadway. Marcus Loew had made a mint with them since his days on Herald Square; the old Broadway Theatre, at Forty-first Street, was showing them. The Astor, one of the city's most elegant legitimate theatres, had gone over to films with *Quo Vadis* on April 19, 1913, for a twenty-two-week run. And the stately Criterion, it had just been revealed, would become a movie theatre after the first of the year, when the Vitagraph Company would reopen it, boasting one of New York's first Mighty Wurlitzer organs (then known as the "Hope-Jones Unit Orchestra") to accompany the pictures, and a stage setting that would reproduce the New York Harbor complete with Statue of Liberty.

The more the Marks thought about it the better the idea seemed. They would attract "high class" crowds to their new theatre (now officially named The Strand) and they would show the wiseacres of Longacre Square a thing or two. Most of the stigma of the storefront peepshow had melted away, as bigger and more respectable theatres were being converted to the showing of films. New York, in 1913, could count 986 movie houses — of all kinds — though even the largest and finest of these suffered from all the architectural ills of the day: tiers of avalanching galleries supported by view-killing columns, hard wooden "opera chairs," and becalmed ventilation systems. They were ill-suited to showing films, and the Marks knew that the new Strand would be perfect.

They had already made free with their hoard of Buffalo nickels, so carefully saved during all the nickelodeon years in upstate New York, to insure that the Strand — to whatever purpose it might eventually be put — was going to be the most beautiful, comfortable and up-to-date theatre in the world.

"From the first moment we conceived of erecting the Strand," stated Mitchell in a newspaper interview, "we made studies of all the best theatres in Europe and America, and we selected Mr. Thomas W. Lamb for the task of putting all their best features into making the Strand a 'National Institution' which would stand for all time as the model of Moving Picture Palaces."

The die was cast. It would be movies for the Strand. But not *just* movies. To expand on this sudden and daring new concept, they started searching for an impresario who would do justice to their ambitions. They didn't have to search long or far; everyone was talking about the sensational accomplishments of young Sam Rothapfel up in Harlem at the Regent. He was obviously the man for the job. On this point, Rothapfel readily agreed, and allowed himself to be lured away from the Regent in spite of the fact that he and a group of associates calling themselves The Photoplay Theatres Company, had just taken a five-year lease on the 116th Street house. When Roxy saw the chance to do his stuff on Broadway, he left the Regent with scarcely a backward look.

When he arrived on Forty-seventh Street, the Marks gave him *carte blanche* to devise the most spectacular entertainment New York had ever seen in a movie house. And Roxy, his head bursting with visions of electric-lit sugar plums, didn't let them down.

The day after the Strand opened (April 11, 1914), Victor Watson, dramatic critic of *The New York Times* wrote:

> Going to the new Strand Theatre last night was very much like going to a Presidential reception, a first night at the opera or the opening of the horse show. It seemed like everyone in town had simultaneously arrived at the conclusion that a visit to the magnificent new movie playhouse was necessary.
>
> I have always tried to keep abreast of the times and be able to look ahead a little way, but I must confess that when I saw the wonderful audience last night in all its costly togs, the one thought that came to my mind was that if anyone had told me two years ago that the time would come when the finest-looking people in town would

be going to the biggest and newest theatre on Broadway for the purpose of seeing motion pictures I would have sent them down to visit my friend, Dr. Minas Gregory at Bellevue Hospital. The doctor runs the city's bughouse, you know.

Among the costly-togged that night were Vincent Astor and his fiancee, Helen Dinsmore Huntington; John Bunny, the beloved fat man of the comedies; George M. Cohan, Sam Harris, Daniel Frohman, Al Erlanger — delegates from Forty-second Street; and William Farnum, who had come to see himself on the Strand's Radium Gold Fibre screen in the opening feature, the first of many productions of Rex Beach's *The Spoilers*.

Roxy had mustered a little army of ushers, doormen and ticket takers, had drilled them in a curious blend of Marine discipline and Chesterfieldian courtesy. The Strand's sumptuous foyer was the scene of much heel-clicking, saluting and bowing from the waist as Old Sergeant Rothapfel's elite guard overwhelmed Broadwayites who had never gotten such attention in a theatre in their lives.

The auditorium, done as a sort of neo-Corinthian temple topped by a vast cove-lit dome, was a revelation. The single balcony swept back in a gentle slope from the loges (fitted out with wicker armchairs) to a broad promenade at the top of the auditorium. The seats both downstairs and up were upholstered in the coziest of Pullman-car plush. Music from a hidden orchestra completed the dream-come-true atmosphere, as the first-nighters took their places for the opening performance in Broadway's first genuine movie palace.

Suddenly three shots of heavy artillery shattered the mood, and the abrupt dousing of the house lights brought on near-panic. But panic was averted in the traditional way as Carl Edouarde led the fifty-piece. Strand Concert Orchestra into "The Star Spangled Banner." In a moment the audience sheepishly realized that the whole thing had been part of the program.

The Moving Picture World reported it this way:

A second afterward we saw in lightning rapidity the scenes which inspired the deathless hymn of glory. In the roar and sweat of battle the starry flag still breezed in eloquent triumph over the brave hearts of the country's defenders. With the swiftness of thought the audience recognized the happy inspiration of sending the new theatre on its career to the strains of patriotic music with this rare glimpse of American glory visualized and, as we all stood up in loving homage, no one failed to congratulate the management on its inspiration. At the same time Rothapfel unmasked his artillery, the darkness in and about the screen was converted into a flood of light and the splendid decorations, the flowers, the hedges of green, the graceful fountains of changing color, and the pretty effects in the wings stood revealed as if by magic.

This horticultural setting for the National Anthem (trade history does not reveal if Rothapfel opened the Family, the Milwaukee Alhambra, the Minneapolis Lyric, or the Regent with similar patriotic fanfare; but it is certain that he launched all his theatre openings for the rest of his career this way) was followed by the *Hungarian Rhapsody No. 2.* Next came The Strand Topical Review, showing baseball scenes made at the Brooklyn Federal League that morning and rushed through the developing fluid to stun the Strand audience that very evening. This was followed by a "scenic" one-reeler made in Italy called *A Neapolitan Incident* and accompanied by a tenor singing "O Sole Mio" behind the screen, then a Keystone comedy which featured the Mutual Girl (public darling of the Mutual Film Corporation) paying a visit to the Strand and being entertained by Roxy himself. Next came the Strand Quartet singing the quartette from *Rigoletto,* which, as Mr. Watson of *The Times* put it, "couldn't have been rendered more perfectly in the great temple of music further down the street."

At last the feature picture, *The Spoilers,* came on the screen — and stayed on for an unprecedented nine reels, each reel lasting fifteen minutes. Not only was Rothapfel's special musical setting much admired; he also was applauded for the happy idea of putting the picture on without any break whatever. "*The Spoilers* is one of those novels we like to read right to the end, if possible," wrote one critic, "and where we cannot indulge in our impulse, we lay the volume down with regret. Breaks and pauses in a running visualization of the novel would be even less welcome than interruptions in the reading. That the audience was well pleased with this new wrinkle was plain. It absorbed the story without an effort and its interest never lagged — at 11:30 we were more interested in the fate of Glenister and all the rest than at 9:15 though we had been looking intently at the screen for more than two hours." Roxy won hearts when he threw out the "One Moment While The Operator Changes Reels" slide and installed four projectors instead.

Only *Variety*, that jaundice-eyed sibyl of show business, failed to predict a long life for the Strand as a "national institution." "The opinion frequently expressed before the Strand opened was repeated after the show people had seen the theatre. It was to the effect that the Strand would not continue with pictures, but would take on a legitimate attraction by next season, probably musical comedy."

Roxy showed *Variety* the error of its ways by topping success with success. "If the other fellow has been giving people a fair picture and an hour's entertainment for a quarter," he disclosed, "I give them a fine picture, and an hour and a quarter's entertainment, and a first-class orchestra to boot for the same price. It's only business, after all. The idea of doing better than the other fellow is the secret of all success."

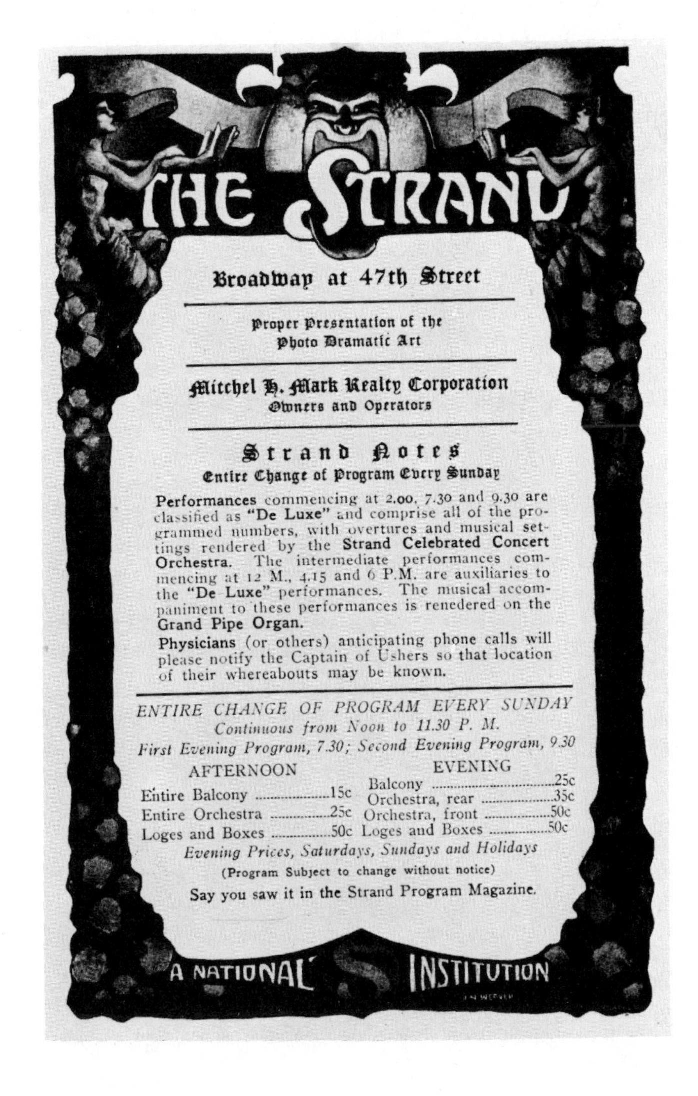

THE STRAND

Broadway at 47th Street

Proper Presentation of the Photo Dramatic Art

Mitchel H. Mark Realty Corporation
Owners and Operators

Strand Notes
Entire Change of Program Every Sunday

Performances commencing at 2.00, 7.30 and 9.30 are classified as **"De Luxe"** and comprise all of the programmed numbers, with overtures and musical settings rendered by the **Strand Celebrated Concert Orchestra.** The intermediate performances commencing at 12 M., 4.15 and 6 P.M. are auxiliaries to the **"De Luxe"** performances. The musical accompaniment to these performances is renedered on the **Grand Pipe Organ.**

Physicians (or others) anticipating phone calls will please notify the Captain of Ushers so that location of their whereabouts may be known.

ENTIRE CHANGE OF PROGRAM EVERY SUNDAY
Continuous from Noon to 11.30 P. M.
First Evening Program, 7.30; Second Evening Program, 9.30

AFTERNOON		EVENING	
Entire Balcony	15c	Balcony	25c
Entire Orchestra	25c	Orchestra, rear	35c
Loges and Boxes	50c	Orchestra, front	50c
		Loges and Boxes	50c

Evening Prices, Saturdays, Sundays and Holidays

(Program Subject to change without notice)

Say you saw it in the Strand Program Magazine.

A NATIONAL INSTITUTION

In 1916, the Rialto's spectacular electric sign was designed by men who knew what light bulbs were for. As the pinwheel showered sparks, the eagle flapped its wings and Old Glory unfurled above Times Square. Present Rialto Theatre on this site is totally new one, dating only from 1932 when original Rialto was razed.

Machine in the World
Co. N.Y.
MACK SENNETT COMEDY
Marie Louise
LADIES HATS
ALL LATEST MODELS
3.50
SPECIALTY IN
GLOVES, HOSIERY, BLOUSES
& UNDERWEAR
Marie Louise

"... A THOUSAND INCANDESCENT LIGHTS"

"Roxy was a showman, and nothing he ever said or did gave the slightest indication that he had an interest in the moving picture . . . except as an element in entertainment. For a long time I considered him the greatest enemy of the moving picture in America, because in his hands the moving picture was falling back almost to its old position as part of a vaudeville program. Yet I can see that in a few respects his influence was beneficial."

GILBERT SELDES

in THE MOVIES COME FROM AMERICA

In spite of the Mark brothers' trail-blazing farther uptown, Times Square — that area just north of Forty-second Street where the confluences of Broadway and Seventh Avenues had begun to eat away the walls of an ever-deepening canyon — still remained the heart of New York's entertainment world in 1915. At its very ventricle, on the corner of Broadway and Forty-second Street, stood a raffish structure called the Victoria Theatre. When old Oscar Hammerstein I (grandfather of the late partner of Richard Rodgers), built the Victoria in 1898 on the site of the Market Livery Stables, he had been in a great hurry. The New York Life Insurance Company had foreclosed the mortgage on his Olympia Theatre across Broadway, and Hammerstein had made a vow that he would have a new theatre built to compete with them before the flag he had hoisted over the Olympia had blown to rags.

On the opening night of the Victoria he stood with newspaper reporters, gazing at the foreclosed Olympia. By Broadway's twinkling light he saw that his flag was still there.

"Now, gentlemen," he said, "you know why I have named my new theatre The Victoria. It is because I have been victorious over mine enemies — those dirty bloodsuckers at New York Life."

The Victoria was a two-gallery house, and Hammerstein lived in the space under the rear of the upper gallery. Here, in two musty rooms, he puttered in his spare time on his "inventions" — a cigar-making machine (on which he held U. S. patent No. 292,958) and some special electric lighting devices. Up on the roof garden of the theatre was a little farm. In this bucolic spot patrons could pay to see a toothsome milkmaid at work on a live cow at certain hours of the day. But, though the Victoria had been billed as "a home for a new form of family entertainment," the rooftop dairy was about as far as it ever went along those lines; the entertainment in the theatre itself was anything but "family."

Managing the Victoria was Oscar's son, Willie, a flamboyant showman. His stock in trade was presenting personalities connected with the more lurid crimes of the day. Among these were Ethel Conrad and Lillie Graham, two maidens who had filled a socially prominent millionaire named Mr. Stokes with a number of bullets, to protect their honors. There was also Florence Carman who had plugged one of her husband's female patients to protect her home and wee ones. Unlike most other Hammerstein attractions, Florence had real talent; exhibiting enlarged photographs of her little brood, she sang "Baby Shoes" until the house was awash. And, of course, there was Evelyn Nesbitt and her red velvet swing.

For some months after the Strand opened, up at Forty-seventh Street, two gentlemen named Crawford Livingston and Felix Kahn had been searching about for a suitable theatre to present the pictures that were being made by their profitable Mutual Film Corporation. None was available in the wave of movie optimism that had swept Broadway, and it became apparent that if Mutual was to have a showcase, Kahn and Livingston would have to build one. But where? Their eyes fell on the Victoria; it was the ideal location, and they were determined to get it. For months they dickered with old Oscar, who was just as determined that they should *not* get it. But finally he gave in. Physically ill from the torture of a sore foot that wouldn't heal, heartsick over the sudden death of his beloved son Willie, and hounded by creditors after the failure of his latest enterprise (the ill-fated Lexington Opera House), he capitulated. Without Willie around to liven things up and keep the Victoria to its tradition of true-life

Grand Guignol spiced with buck-and-wing, Oscar didn't see much point in hanging on. He needed the money, and the Victoria was sitting on the most expensive piece of theatrical real estate in the world.

Once the deal was set, Kahn and Livingston hired Roxy away from the Strand with the promise of a princely $200 a week and a free hand in making the new theatre another Rothapfel landmark. The result was the Rialto Theatre . . . "Temple of the Motion Picture — Shrine of Music and the Allied Arts."

But Oscar was not quite so easy to deal with as Roxy was. Having no notion of giving up his franchise on the corner of Forty-second and Broadway, he leased office space in the forthcoming Rialto building even before the wreckers' bars had started to dismember the Victoria. Once assured that he could come back to his old stand again, he refused to vacate his dingy digs under the balcony until he was literally torn out with the plumbing. Then he set up temporary quarters in the Republic Theatre (which he also owned) next door on Forty-second Street, finally moving across the street to a room in an old building, where he kept his vigil as the Victoria went down and the Rialto went up.

The Rialto went up slowly — too slowly to suit Roxy and Crawford Livingston, and much too slowly to suit Hammerstein who was impatient to move into his new office. The original opening date had been set for April 1, 1916; on April 4, with the opening still weeks away, Hammerstein could contain his impatience no longer and stormed the battlements, determined to move into his new office on the date it was promised, ready or not. He was summarily thrown out by workmen on the job, and went next to the West Forty-seventh Street Police Station — that bastion of show-biz law and order — to seek protection from the bodily harm he was convinced the Rialto huskies meant him.

Roxy, when questioned about the matter, stated that Mr. Hammerstein had been "politely but firmly refused admission to the building because it was feared that he might make a speech to the workmen and delay the completion of the theatre, now months overdue."

The New York Times continued the story, headlined "Oscar in Three Reels," after a brief synopsis of "what has gone before":

Reel II. Hammerstein returned to the Rialto yesterday and found a barricade had been built. Immediately he went to the West Side Court and obtained a summons for Crawford Livingston, President of the Rialto Corporation, and Robert E. Hall, superintendent of the new building. Mr. Hammerstein wants to know why there were padlocks and other things intended to discourage him when he went to the Rialto yesterday—and the matter will be discussed in court this morning.

Reel III. Later in the day yesterday, Mr. Hammerstein, nothing daunted, and recalling the adage about trying again, returned to the theatre. He could not gain entrance by the door, so he scaled the fire escape and soon came upon Mr. Hall who was lying in wait. There were words, many of them, and Patrolman Drecker, happening to pass by and overhearing the conversation, suggested that both gentlemen accompany him to West Forty-seventh Street to tell their troubles to Captain Underhill. The Captain suggested that they call an armistice until court convenes in the morning.

S. L. Rothapfel, director of the Rialto, stole the last scene in the episode. "Mr. Hammerstein's little affair," he said, "is jeopardizing our opening which has been set for a week from Saturday. That is his loss as well as ours, for I had him on the invitation list and was arranging a little complimentary feature in his honor in consideration of his long managerial activities on that corner. If he continues to give an imitation of Mt. Vesuvius in eruption, we may have to withdraw the invitation out of consideration for our other guests on that occasion."

The Rialto opened on Good Friday — April 21, 1916 — with "the peal of the grand organ, the fanfare of the orchestra, and the flash of a thousand incandescent lights . . ." but without Oscar Hammerstein.

The Rialto's First Anniversary souvenir program (see following pages) showed the wonders of the theatre and its bugle-and-baton-bearing staff. It also underlined Roxy's belief in the importance of music, listing the ambitious "music appreciation" course he gave fans as part of the first year's curriculum.

THE RIALTO

"TEMPLE OF THE MOTION-PICTURE"

SHRINE OF MUSIC AND THE ALLIED ARTS

BROADWAY & 42nd STREET, NEW YORK

PERSONAL DIRECTION OF S. L. ROTHAPFEL

THE FOYER

THE UNIFORMED STAFF

THE DOME

HUGO RIESENFELD

The Musical Director

HUGO RIESENFELD, the masterly conductor of The Rialto Orchestra, was born in Vienna. He graduated from the Vienna Conservatory, and, after leaving that celebrated institution, became a concert violinist. When Gustave Mahler, director of the Vienna Opera House, heard him play he engaged him at once and it was at the Imperial Opera House that Dr. Riesenfeld conducted the Ballet and produced the first ballet which he composed himself. It was called "Chopin's Dance" and met with enthusiastic approval.

Oscar Hammerstein brought Dr. Riesenfeld to New York and for four years he was engaged at the Grand Opera House as concert master and conductor. It was while there that he won the hearty commendation of the New York musical critics by his rendition of the "Meditation" from "Thais." This performance alone would have stamped him as one of the most sympathetic masters of the violin in America. Later Klaw and Erlanger engaged him as conductor and produced a comic opera which he composed.

Mr. Rothapfel, hearing Riesenfeld interpret a Strauss waltz, engaged him as conductor at The Rialto, where he has had charge of the orchestra ever since the house opened. By virtue of his true musicianship, his unflagging enthusiasm, his grace, and his personal charm he has made himself perhaps the most widely popular musical director in the city. Upon occasion he conducts after the Continental fashion, playing the violin and leading the orchestra at the same time. His interest is not confined to the musical programme proper. He is equally sincere in his musical interpretation of what is transpiring on the screen and is a thoroughly able exponent of Mr. Rothapfel's theories along that line.

THE ORCHESTRA

The Music

IT is impossible to ascertain just what percentage of The Rialto's patrons are attracted here primarily by the music, but unquestionably a goodly proportion of them get as much entertainment from the orchestra and the soloists as they do from the pictures. Mr. Rothapfel originated the idea of presenting motion pictures in conjunction with a high class musical programme and in The Rialto he has developed that form of entertainment to a point as yet unequaled. In point of size the orchestra here is exceeded only by that in the Metropolitan Opera House. In the quality of its work it is equal to the best. With a forty piece orchestra and a grand pipe organ—the largest yet installed in any theatre—it is possible to get the full effect of the most ambitious symphonic compositions, and when this organization is taken in conjunction with the appearance of artists who have established reputations in the fields of opera and concert, the musical programmes reach a standard of quality quite unattainable elsewhere for anything like the same price of admission.

In addition to the musical programme proper, each pictorial offering on the bill has its own incidental music, painstakingly selected and arranged during the week prior to its presentation. In the case of the so-called feature pictures which form the dramatic attraction of each performance, this incidental music has become of such importance that many a picture of merely average merit has become lifted into widespread popularity by the heightened effect which Mr. Rothapfel's musical setting has given it. Each scene in the story is considered for its emotional qualities and those qualities are paralleled as closely as possible in the mood of the music which accompanies the pictures. Where nothing in the vast musical library of The Rialto seems to fit the scene precisely, someone or other—there are seven recognized composers in the Rialto Orchestra—composes a theme which expresses the spirit of the story at that particular point. Effects have been produced by this interpretative music which were nothing short of electrical and the idea of interpretative music for the pictures, as exemplified at The Rialto, has been adopted wherever pictures are shown.

The full orchestra is heard at four performances daily. The intermediary performances are interpreted on the grand organ. The Rialto boasts as its organists Dr. A. G. Robyn, composer of "The Yankee Consul," "All For The Ladies," and other light opera successes, and Prof. Firmin Swinnen, also a composer and formerly organist of the noted Cathedral at Antwerp.

As an indication of the standard maintained by this institution in the music it presents, a list is given on a subsequent page containing the names of artists who have appeared here, together with the overtures which have been rendered by the orchestra.

It is a fact worthy of note that, when considered during an equal period of time, no other institution in the world, neither opera house nor symphony hall, possesses the musical scope represented by this list.

Artists Who Have Appeared At The Rialto

SENOR VICENTE BALLESTER, Baritone, Boston Opera Company.
MME. JEANNE MAUBOURG, Mezzo Soprano, formerly Metropolitan Opera Company.
M. DESERE DeFRERE, Baritone, Chicago Opera Company.
ALBERT GREGOROWICH JANPOLSKI, Basso, Boston Symphony Orchestra.
SIGNOR MARION RODOLFO, Tenor, San Francisco Opera Company.
MARY BALL, Soprano.
ANNA MURRAY, Contralto.
HENRY MILLER, Basso.
ALFRED DeMANBY, Baritone.
SASCHA FIDELMAN, Violinist.
GASTON DUBOIS, Cellist.
SASCHA JACOBSEN, Violinist.
VINCENT BACH, Trumpeter.
MME. JENNY DUFAU, Soprano, Chicago Opera Company.
MLLE. MADELEINE D'ESPINOY, Soprano, Opera Comique, Paris.
COUNT LORRIE GRIMALDI, Basso, formerly Metropolitan Opera Company.
MME. REGINA VICARINO, Soprano, formerly Manhattan Opera Company.
ELSA DIEMER, Soprano.
HILDA GOODWIN, Soprano.
JAMES PRICE, Tenor.
AMPARITA FARRAR, Soprano.
HELEN JEFFREY, Violinist.
HANS KRONOLD, Cellist.
BELA NYARY, Cimbalum Soloist.
A. KASTNER, Harpist.
FRANK CORK, Xylophone Soloist.

Rialto Male Quartette

JOHN YOUNG, Tenor
GEORGE REARDON, Baritone
HORATIO RENCH, Tenor
DONALD CHALMERS, Basso

Overtures Which Have Been Played By The Rialto Orchestra

OVERTURE	COMPOSER
"Night in Granada"	C. Kreutzer
"Hungarian Rhapsody XIV"	Liszt
"Irish Rhapsody"	Victor Herbert
"Zampa"	Herold
"Prelude"	Liszt
"Southern Rhapsody"	Lucius Hosmer
"Oberon"	Weber
Selections from "Tosca"	Puccini
"Vi Rhapsodie"	Liszt
"Sicilian Vespers"	Verdi
Ballet Suite from "Queen of Sheba"	Goldmark
"Carneval"	Dvorak
"Merry Wives of Windsor"	Nicolai
"Patrie"	Bizet
"The Flying Dutchman"	Wagner
"Die Schone Galathea"	Von Suppe
Selections from "Aida"	Verdi
"La Forza Del Destino"	Verdi
"Sakuntala"	Goldmark
"William Tell"	Rossini
"Phedre"	Massenet
"1914" (Descriptive Overture)	Hugo Riesenfeld
"Der Meistersinger"	R. Wagner

OVERTURE	COMPOSER
"Bacchanale," from Samson and Delilah	Saint Saens
"First Hungarian Rhapsody"	Liszt
"Robespierre"	Litolff
"Rhapsody Espana"	E. Chabrier
Vorspiel Act III "Lohengrin"	R. Wagner
"March Slav"	Tschaikowsky
"Fledermaus"	Johann Strauss
"Hermit's Bell"	Maillard
"If I Were King"	Adam
"Stradella"	Flotow
"Raymond"	Thomas
"Martha"	Flotow
"Orpheus"	Offenbach
"Dance of The Hours," From "Gioconda"	Ponchielli
"Rienzi"	R. Wagner
"1812" (Descriptive)	Tschaikowsky
"Hungarian Rhapsody No. 2"	Liszt
"Il Guarany"	Gomez
"La Boheme"	Puccini
"Tannhauser"	Wagner
"Cricket On The Hearth"	Goldmark
"Madam Butterfly"	Puccini
"Mignon"	Thomas

The Times called it "a new palace of polite pleasure for the thousands." Thomas W. Lamb, well on his way to becoming court architect for Rothapfel houses, had designed the new pleasure palace, and while its handsome Adam interior was not nearly so vast as the Strand's, it was every bit as elegant and ever so polite. So were the Rialto ushers, and their uniforms were even more gorgeous than their manners.

They wore scarlet tunics piped in gold and looped across the front with more gold and tassels. Each of them carried a swagger stick with mother-of-pearl tips that lit up in the dark. The Head Usher carried a bugle. It is not known what the bugle was used for unless it was to rally the troops in time of riot; but the swagger sticks had a definite purpose beyond that of directing guests to their seats in the dimness. The Rialto ushers were all experts in first aid, and the swagger sticks (as a Rothapfel press release revealed, deadpan) were to be used in making tourniquets.

Thus prepared for panic and bloodshed, the Rialto opened its doors. Opening-day ads proclaimed "The World's Largest Grand Organ — Superb Concert Orchestra — Most Wonderful System of Electrical Effects Ever Installed in Any Theatre — 15¢ — 25¢ — 50¢ — No Higher."

Outdazzling all other guests at the Rialto's opening was Mary Pickford. With her were her mother and Adolph Zukor, who knew that beneath those golden broomstick curls lay the shrewdest little brain in show business; he had already been pile-driven into paying her $10,000 a week — not as salary but merely as guarantee against the 50 per cent share of the profits from all her pictures that Mary demanded. As America's Sweetheart recalls in her autobiography, *Sunshine and Shadow*, Zukor once remarked: "Mary, sweetheart, I don't have to diet. Every time I talk over a new contract with you and your mother, I lose ten pounds." Though protesting that she really didn't want to, Mary was constantly threatening to decamp to Universal or Triangle or whoever was making the best offer at the time, and Zukor rarely let her out of his sight lest some unscrupulous producer make off with the Famous Players' bread and butter.

Marcus Loew and B. A. Rolfe—rival exhibitors on the Broadway turf — came to the Rialto opening night together, united in a common rivalry stronger than their individual ones. Rolfe had inherited Roxy's job — if not his crown — at the Strand; Loew, with a big chain of movie houses, was just beginning to fancy himself as king of Broadway.

At eight-fifty a trumpet call announced the beginning of the entertainment and Alfred Robyn entered from the side of the auditorium and took his place at the white console of the big Austin organ. He tapped a bell twice and the musicians in dark red tuxedo jackets filed into the orchestra enclosure and were seated. Then Hugo Reisenfeld, conductor of the orchestra, arrived, carrying his violin with him. With a wave of his bow, the orchestra struck up "The Star Spangled Banner." Soon they were reinforced by sixteen boys in white duck sailor suits, and then the full complement of Rialto vocal soloists, male and female. At the conclusion of the overture, which Reisenfeld conducted in the Continental fashion while playing his violin, he dragged (or seemed to drag) Roxy out before the audience. Roxy acknowledged the applause with dignified genuflections and was gone without wasting any time on speeches.

After a "Topical Digest of News" which featured the Rialto ushers drilling on the roof, Mary Ball appeared on one of the platforms flanking the screen and sang two songs, accompanied on the organ by Dr. Robyn. She was followed by Violet Marcella on the opposite platform who performed a brief classical dance — though, as *Variety* commented, none too brief. Then came a Prizma colored scenic film, a visit to that other Rialto — the one in Venice. Next, First Concertmaster Sascha Fidelman rendered Sarasate's "Zigeunerweisen" as a solo. The feature picture followed — Douglas Fairbanks in a western called *The Good Bad Man*, a Triangle Picture. As an after *piéce de résistance* Roxy presented the noted baritone, Alfred De-Manby, who sang two songs and, as an encore, joined the Rialto Quartette to render "a special arrangement of a neat little musical comedy number." The show closed with a Keystone comedy,

The Other Man, featuring Fatty Arbuckle in a dual role and accompanied on the organ by Professor Firmin Swinnen who played a two-reel version of *Til Eulenspiegel's Merry Pranks.*

In pointing out some of the unique features of the Rialto (the spectacular electric sign that seemed to whirl like a pinwheel into a shower of sparks that spelled out R-I-A-L-T-O as an eagle flapped its wings and Old Glory shone above; the homelike overstuffed furniture in the lobbies; the writing desks and retiring rooms on the mezzanine; the filtered Puro drinking water; the carriage call in the main lobby; the banks of fresh flowers everywhere) *The New York Times* discovered the most unique thing of all about the Rialto: "The Knickerbocker is a fine old theatre temporarily made over into a movie house, and even the Strand is built so that at very short notice it could be converted to the uses of opera or drama, but the Rialto is a motion picture house pure and simple. It is stageless, the screen being placed boldly against the back wall of the theatre. It is built in the conviction that the American passion of the movies is here to stay."

Behind the proscenium arch was a classic colonnade supporting a half-dome. Beneath this dome, in the space where a stage apron would normally have been, sat the Rialto Orchestra. (See page 48.) This did not mean that Roxy had forsaken the idea of soloists in his presentations. The openings in the colonnade on either side of the screen revealed elaborate stage settings in the neo-Belasco–Maxfield Parrish tradition, when the red-plush curtains were drawn back. These side stages were always decorated to carry out the theme of the photoplay on the Rialto screen, and here Roxy's soloists performed.

Banks of lights in various colors were concealed in the ledges around the dome, behind columns, back of translucent panels set in the walls and ceiling. These were controlled by dimmers so that the colors could be changed gradually. During a movie each scene had its own color scheme as well as its own music, and Roxy's "color harmonies" became one of the Rialto's most talked-about features.

Soon after the theatre opened, Hugo Reisenfeld received an honorary doctorate — in a typically Rothapfellian fashion. Noting that the Rialto program listed "Professor Firmin Swinnen at the Largest Organ Yet Installed in Any Theatre," Roxy sent for Reisenfeld. "Have you ever been to college?" he asked the conductor.

"Yes," replied Reisenfeld. "Some."

"For instance?"

"For instance the Vienna Conservatory."

"Very good, Hugo. That's all."

Roxy turned to his press agent (none other than Terry Ramsaye, later author of the first comprehensive history of the movies, *A Thousand and One Nights)* and said, "No organist is going to outrank my musical director. From now on he'll be Doctor Reisenfeld. See?"

Next week's program noted the presence of the new doctor in the house, and the title stuck with him — to his not-too-keen discomfort — the rest of his days. Legend has it that performers at the Rialto used to drop into Dr. Reisenfeld's office and ask for medical advice. They got advice, but it wasn't medical.

Music was the key to Roxy's showmanship. And though he responded to it with an emotional extravagance that could make every fibre in his stocky five-foot-eight frame vibrate like an aeolian harp, he couldn't read a note. One night, he decided that some of the glory that shone on Dr. Reisenfeld should be Roxy's. Into the pit he went, wearing a maroon velvet jacket that was the mark of Rialto musicianship, and took the baton to conduct the "1812" Overture. He finished a good four beats ahead of the orchestra (or, as a Broadway wag put it, somewhere in 1808) and left the pit considerably less embarrassed than the musicians.

In the meantime the Rialto Theatre Corporation under Livingston and Kahn, had leased the site of the old Palmer-Singer garage, up at Broadway and Forty-ninth Street, from Maurice Heckscher, to build the new Rivoli Theatre. The new management decided to make Roxy the director of both theatres — two impresarios for the price of one Roxy.

This bronze plaque was later installed in a niche on the Grand Staircase of the Roxy Rotunda; is now in the Museum of the City of New York, the gift of the Rothafel family.

Broadway Gets A Parthenon

The Rivoli opened on December 28, 1917, with Douglas Fairbanks in *A Modern Musketeer* and a platoon of Roxy soloists, patriotic tableaux, organ, and orchestra. Designed by Thomas W. Lamb, "The Triumph of the Motion Picture" had a classic façade — the marquee was surmounted by a row of white terra-cotta columns topped by a pediment right off the Parthenon. The Rialto (The Temple of the Motion Picture) had a front built of tapestry brick to support its flashy electric pin-wheel sign. It was only natural, then, that the Rivoli's presentations (as well as its façade) be somewhat more high hat than the slightly *gemütlich* offerings of the Rialto. In this case, however, class didn't seem to pay off; Roxy's budget at the Rialto, for a typical week, was around $9,000 with a gross of between $19,000 and $20,000. The Rivoli operated at a cost of $13,000 a week and its profits hardly showed at all.

But as long as it didn't lose money, the Rivoli was worth its weight in overhead as a prestige operation. And prestige it did surely have. A characteristic presentation was the one for the week of November 3rd, 1918. The Rivoli program noted that "the material comprising the programs at both the Rivoli and the Rialto is personally selected by Mr. Rothapfel, who also supervises the staging and lighting of the music numbers." And a sharp eye would also note that many of the artists playing the Rivoli must have shuttled up Broadway, still in greasepaint, from the Rialto at Times Square.

The bill opened with the Rialto Male Quartet singing (perhaps a bit breathlessly) "Love's Old Sweet Song." This was followed by the overture, von Suppe's "Beautiful Galatea," played by the Rivoli Orchestra. (Dr. Reisenfeld was nominally *director* of the orchestra, but the conductor was a man whose destiny would be caught up with Roxy's from then on, the fire-breathing little Hungarian, Erno Rapee.) Then came an educational novelty on the screen, *Nature's Mischief Makers*, followed by a Van Scoy Scenic Study, *Outdoors With Beauty*, both accompanied by the orchestra. Next appeared another name whose career would be in Roxy's hands for years to come: Gladys Rice, the soprano, singing, "Oh, Dry Those Tears"

Thomas Lamb's Parthenon-inspired façade graced the Rivoli programs in 1917. The ubiquitous "R's" stood for many things—Rothapfel, Rivoli, Rapee, Rialto, Regent or Reisenfeld—but never Ragtime. Today Rivoli is gone, demolished for an office block.

accompanied by Sepp Morscher on harp, William Feder on cello, and Professor Firmin Swinnen at the organ. *The Creel Official War Review* (one of a series of government-produced films for which Roxy had served as technical adviser) came next, and then the Rialto Male Quartet reappeared to set the mood for the feature picture by singing "My Own United States."

The picture, *Safe for Democracy*, was a J. Stuart Blackton presentation "founded on General Crowder's 'Work or Fight' order," and starred Mitchell Lewis as Big Steve Reardon with a cast of characters that included a Shipyard Superintendent and an Agitator. The show closed with Professor Swinnen playing the Toccata from Widor's Fifth Symphony — a tour de force that would have made organists of the later Mighty Wurlitzer era blanch in terror.

Patrons were advised that performances at two-fifteen, seven-thirty and nine-thirty would be "full presentation performances, with the Rivoli Orchestra, soloists, color harmonies and scenic effects. The musical setting for the intermediary performances would be supplied by the largest pipe organ yet installed in any theatre."

Roxy at this time was drawing a salary of between $500 and $700 a week, depending on the box office percentages of the Rivoli and the Rialto. He was enjoying his dual job considerably, and decided to move his headquarters from the Rialto to the slightly more swanky Rivoli uptown. He furnished his new office in elegant style, and was somewhat put out when he learned that the corporation's purchasing agent had no idea of paying for the new furniture. Roxy was establishing a pattern of personal grandeur that was to grow and grow (in spite of a few small-town touches he never lost). He had a Japanese body-servant who hovered about the office to respond to the first snap of the Roxy fingers; few people knew that this same Broadway samurai was a whiz at frying hot dogs "Forest City style" on a hot plate in the adjoining bathroom.

Roxy's ideas of splendor had for some time been cause for alarm to the Rivoli-Rialto operators. There were monthly fights over the cost sheets. Finally, when the auditors submitted that Roxy ought to pay for the long distance phone calls between the Rivoli office and the race track at Havre de Grace, there was open war.

Roxy was ready. He was fresh from a short tour of duty with his beloved Marines, and had been made a major USMC Reserve, for his contribution to the *Creel Official War Review* films. While his habit of affecting a flowing cape with his uniform brought grumblings from Marine high brass, he had accomplished his wartime job in grand style, and he had returned to the Rivoli more spit-and-polish than ever. So, when the management's financial corporals had the temerity to bring up the matter of the race-track phone calls, Major Rothapfel counterattacked in the most devastating way he knew: he resigned.

Hugo Reisenfeld was named his successor at the two theatres, and Roxy, after borrowing $400 from Henry Heil, the doorman at the Rialto, and selling his office furniture to Hugo at the Rivoli, embarked on one of the most curious enterprises of his career.

Frank G. Hall, a New Jersey promoter, approached Rothapfel with a proposition that seemed the very antithesis of all Roxy had worked so hard to accomplish in the ten years he had been guiding the destiny of picture-theatre presentation. Hall proposed that they set about filming and distributing the "Rothapfel Unit Programme" — a complete bill of canned entertainment for film theatres. The appeal to Roxy lay in the fact that it was a chance for him to make his ideas and fancies available to theatres which could not afford live presentations on even the most modest scale. Terry Ramsaye recalls that once, when he went to visit Roxy at the studios where a unit was being filmed, he found Roxy, with an eighteen-piece orchestra playing, busily directing a cameraman on how to make a closeup of a clock about to strike midnight — and no one else in the studio. "That," Ramsaye observes, "was mood. For Hollywood it would have been commonplace. For Roxy it was indicative."

The Rothapfel Unit Programme was launched with enormous fanfare. On May 19, 1918, he unveiled the first Programme, for exhibitors only, at the Rialto — a news magazine reel, a one-reel novelty prelude, a five-reel feature, a pictured epigram and a one-reel comedy. That night a big banquet was given at the Astor Hotel. Present, besides the captains of the movie industry and a squad of Marine colonels, there were newspaper people imported from all over the country. A trade magazine account of the banquet noted that "there were many speeches by representative men in the trade, and Miss Louella Parsons, of the New York *Morning Telegraph,* upheld the dignity of the women in making a sensible and clever speech when called upon."

In spite of the lovingly phrased blurbs that filled the opening announcement ads for the Rothapfel Unit Programme (sample: "Rothapfel has done for the Movies what Wagner did for Opera; he has coordinated the Arts into one harmonious whole." — Sylvester Rawling, New York *Evening World)* the project turned out to be more than a bore; it was an albatross. Few exhibitors signed up, and with the showing of Unit Programme Number One at the Cosmopolitan Theatre on Columbus Circle (William Randolph Hearst's 1,400-seat token of his friendship for Marion Davies, and a house notable for its mortality rate) the whole idea shriveled on the vine. Roxy's partnership with Frank Hall was dissolved, and both went their separate ways.

Bowes Knows!
REALART PICTURES
CAPITOL
CAPITOL
CAPITOL
Edward J. Bowes

FROM WHITE ELEPHANT TO GOLDEN GOOSE

"As if the very hugeness and magnificence of the new Capitol Theatre were not enough to stagger the imagination, the spectacle presented there last evening by the Messrs. Bowes and Wayburn leaves the mind of the beholder reeling. As one sat amongst that vast and festive throng one felt somehow very humble that all this pomp had been decreed for the edification of mortal man."

NEW YORK SUN,
October 25, 1919

The Capitol Theatre was the brain child of Messmore Kendall, lawyer, boulevardier, and Son of The American Revolution. At the end of World War I, he leased a parcel of land from the Jacob Wendell estate, on the corner of Broadway and Fifty-first Street, as a real estate investment. It was occupied by a livery stable of Augean proportions, a blacksmith shop, and (symbol of changing times) a filling station. But there was an eccentricity in the lease: no part of the site was to be sublet to any concern engaged in the manufacture of cosmetics or corsets. Fortunately it did not preclude the use of these items on the premises—and Kendall figured it would be an ideal spot for a theatre. Not just any theatre, though; he dreamed of a motion-picture palace so vast and so splendid that it would make the Strand and the Rialto and the Rivoli look like peepshow parlors.

He interested a group of gentlemen of widely varied backgrounds to join him in the enterprise. Forming the board of directors of the Moredall Realty Corporation, in addition to Messmore Kendall (whose name had inspired the "Moredall" title) were General T. Coleman Du Pont (a flourishing Daddy Warbucks who had been disappointed in his ambition for a Presidential nomination in 1916); former U. S. Postmaster General Frank H. Hitchcock; George H. Doran, the publisher; William Braden, the copper magnate; Georges Armsby, a wealthy California packer; Robert W. Chambers, author of the 1906 bestseller, *The Tracer of Lost Persons*, which gave immortality to the indefatigable Mr. Keene; and Major Edward J. Bowes.

The Major (he had recently served as a staff specialist in Intelligence—hence the title, duly come by and never relinquished) was the self-educated son of an Irish dock weigher in San Francisco. He had lost one fortune in the Great Fire of 1906, made another by prophesying where the city's business district would rebuild. After a period as a civic reformer (the Tong Wars of San Francisco's Chinatown were his target), he came East and became interested in theatrical real estate, including The Cort Theatre on New York's Forty-eighth Street, a legitimate house. During World War I he served in the Army, and the Armistice found him eager to get back to the business of making money. When Messmore Kendall suggested that he join him and Du Pont in the Capitol Theatre venture, the Major was delighted.

They retained Thomas W. Lamb, designer of the Regent, the Strand, the Rialto and the Rivoli, to draw up plans for what was to be the largest theatre in the world, with 5,300 seats and an elegance never before seen by man. For the Capitol, Lamb created an auditorium of magnificent distances. The style was Adam and Empire, with fluted columns, damask hangings, and a great dome decorated with classic bas-reliefs. Instead of the usual gold leaf, Lamb lavished silver leaf on almost everything, with stunning effect. In the lobby, clearly visible from the street, was a sweeping staircase of snowy marble, scrubbed nightly by a team of dedicated charwomen imported, according to legend, from Baltimore.

The lobby itself was paneled in mahogany (a fact that postponed the opening of the theatre for several days until the New York Building Department, armed with blow torches, was satisfied that everything was sufficiently fireproof). Murals by William Cotton rose above the paneling, and three gleaming rock-crystal chandeliers lit the scene. These were part of a set of sixteen that Major Bowes had bought from Sherry's for $75,000 when that fondly remembered landmark on Fifth Avenue was demolished shortly before. In the Capitol they added considerably to the atmosphere of Federal *bon ton* set by the name of the theatre.

"Capitol" might possibly have been General Du Pont's idea—a monument to his thwarted political hopes. Be that as it may, the whole place was just the proper blend of Broadway and Connecticut Avenue to please everybody.

The Capitol building took up nearly the entire blockfront between Fiftieth and Fifty-first Streets, and was faced with white vitreous brick. Steel for the structure had presented a problem, due to wartime restrictions, but the difficulty was solved when the builders contracted for all the heavy trusses that had been used to support Broadway while the Interborough Rapid Transit was being built. Girders for the theatre building were fabricated from the IRT steel right on the spot; nothing could have been more convenient.

The top floor of the six-story office building that fronted on Broadway was given over to a baronial twelve-room apartment occupied by Messmore Kendall and his wives (successively) until his death forty years later. The drawing room was a Tudor extravaganza reminiscent of the Watching Chamber at Hampton Court; it was complete with mullioned windows of stained glass, a great stone mantel, and a secret panel. Behind the panel lay a winding staircase that led down into a sort of upholstered eagle's nest. This was fitted out with leather sofas facing a wall of plate glass; on the other side of the glass lay the cavernous auditorium of the Capitol. The window could be lowered on occasion, and to this "watching chamber" all his own, Kendall would lead guests after dinner to see the show taking place on the stage a whole city block away.

The apartment below Kendall's was the domain of Major Bowes and his beautiful wife, Margaret Illington, the former Mrs. Daniel Frohman. What the Bowes' apartment lacked in secret panels it made up for in *objets d'art*; there were enough oil paintings, ormolu jardinieres and oriental statues to stock a dozen auction galleries. They entertained extensively, and on important occasions the Major himself would go into the kitchen overlooking

The Capitol Lobby gave New York its grandest staircase in 1919. The French rock crystal chandeliers were three of a set of 16 salvaged from the old Sherry's on Fifth Avenue by Major Bowes.

Broadway and prepare his *specialité*—a pot of baked beans.

The Capitol Theatre opened at 7:00 P.M. on the evening of October 24, 1919. Passers-by on Fiftieth Street near the theatre's stage door that morning had witnessed a curious scene: there were the directors of the Moredall Realty Corporation busily mixing mortar. What appeared to be a cornerstone ceremony was nothing of the sort. A bricklayers' strike a few days before had left the Capitol's backstage wall with a yawning hole in it, and last-minute negotiations with the union had failed to get the masons back on the job. So Messmore Kendall had called an early-morning board meeting, had distributed aprons and trowels, and had told his high-salaried scab laborers to get cracking.

The opening that night was brilliant, both socially and politically if not artistically. There was a reception for invited guests in the vaulted foyer at the top of the white marble stairs with a thousand jostling emissaries from the worlds of the theatre, government, industry, and the military present. At eight o'clock gongs were struck (a sound as dear to the Major then as in later years) and everyone was herded into the auditorium where the paying audience had already gathered for the main event of the evening. And a long evening it was.

First, Ernest E. Jores performed a Concert Overture in C Major on the big Estey pipe organ with illuminated push-button stops. The organ console was in the center of the orchestra pit, the rest of which was covered over with an imitation smilax shroud. Up on the stage was sprawled Arthur Pryor's seventy-piece concert band, and behind them was another smaller stage, where the divertissement and the films were to be presented.

A newsreel followed the organ overture; then a tone poem, "After Sunset," by the band; next Miss Lucille Chalfant rendered Gounod's "Mireille," followed by a Prizma color picture. Then Arthur Pryor himself appeared to conduct the band in the overture to *I Promessi Sposi*, the *Grand Scene from Andre Chenier*, and "The Capitol March," composed specially for the occasion. Next came a Travelaugh put together by Hy Mayer, the Capitol's film editor and photographed by the Capitol's cameraman, James Prangley. This was followed by a Universal comedy, *The Eternal Triangle*, performed by some dogs. Next on the screen came scenes of Ned Wayburn, former Ziegfeld producer, rehearsing the evening's *pièce de resistance*, the *Demi Tasse Revue*.

Finally the curtains opened on the *Demi Tasse Revue* itself — a drip-pot-pourri in eleven acts. Here's the way the program ran:

1. "You're The Finest Of Them All," *sung by* Paul Frawley and Lucille Chalfant
2. "The Story Book Ball," *sung by* Muriel De Forrest
3. "Milady's Dressing Table"
   POWDER PUFF................Dorothy Miller
   ROUGE ....Janet Starr
   MOTH ....Pearl Regay
   CANDLE ....Paul Frawley
4. "A Story Dance," *by* "Jim" Toney
5. "Shadowland" . . . "A Dainty Act Altogether"
   "JUST FOR ME AND MARY," *sung by* Paul Frawley and Lucille Chalfant
   "SILHOUETTES," *danced by* the Capitol Ballet
   "SWANEE," *sung by* Muriel De Forrest
6. "By the Fire Light"
   "HOW CAN YOU TELL," *sung by* Paul Frawley
   Assisted by A Girl, Her Suitor, Her Father, Another Suitor, The Singing Conductor, and the Capitol Ballet
7. "Indian Summer," *sung by* Lucille Chalfant
8. "Laughing Waters," *sung by* Mae West
9. "In Arizona," *sung by* Will Crutchfield
   Assisted by Ranch Girls, Rope Spinners
   Rope-Spinning Specialty—Will Crutchfield
10. "Vampires"
   "OH, WHAT A MOANIN' MAN," *sung and danced by* Mae West, Arthur Franklin at the piano
   "SPECIALTY" "Jim" Toney and Ann Norman
11. "The Capitol Tower"
   "UNDERNEATH THE HONEY MOON," *sung by* Muriel De Forrest
   "EVENING STAR"—Pearl Regay
   "COME TO THE MOON"—Finale
   MAN IN THE MOON..........Paul Frawley
   GIRL IN THE MOON........Lucille Chalfant
   THE TWINKLE STAR GIRLS, THE MOON BOYS

Major Bowes put Arthur Pryor's string-augmented band on stage à la Roxy while *Demi Tasse Revue* performed on small stage-within-a-stage at rear.

Ned Wayburn's *Demi Tasse Revue* featured acts like "Milady's Dressing Table" (upper right) in which Rouge and Powder Puff looked on knowingly while Moth flitted around Flame. Paul Frawley and Lucille Chalfant (below) were the Man and the Girl in the Moon, and Muriel De Forest (above) introduced a song called "Swannee" by G. Gershwin. Mae West (lower right) was a very naughty vampire.

Finally, at 11:20 P.M. the feature picture went on, for those who could still hold their eyes open. It was called *His Majesty, The American* and starred Douglas Fairbanks in the world's first United Artists production.

When George Gershwin had heard that Ned Wayburn was producing the opening show at the new Capitol Theatre, he decided to write some music for the event. He and Irving Caesar composed the tune and lyrics for one song in an afternoon; the other was a melody George wrote without any words. Wayburn bought both numbers, but it was the wordless tune that he liked best. With lyrics by Wayburn and Lou Paley, it emerged as "Come to the Moon"—the finale song for the *Demi Tasse Revue.* The set was a double spiral staircase which revolved as Twinkle Star Girls danced with electric lights on their slippers and the Moon Boys sang. With the rest of the stage in darkness, it was quite a sight. The other Gershwin song, the one that Irving Caesar wrote the words to, was called "Swanee."

"Swanee," as rendered by Pryor's brass band and Miss De Forrest, wasn't much of a hit. As a matter of fact, when George and Irving loitered in the lobby after the show to watch how the sheet music was selling, they were mortified to see how few buyers there were for either number. Weeks afterward, Al Jolson heard Gershwin play "Swanee" at a party and insisted on interpolating it among the Romberg items in his revue, *Sinbad,* which was already playing across the street from the Capitol at the Winter Garden. "Come to the Moon" seems to have been doused along with the electric ballet slippers when the *Demi Tasse Revue* came to the last drop. But "Swanee" keeps on rolling along.

It didn't take long for Major Bowes and his associates to realize they had too much of a good thing. The critics were outspoken in their comments on "unwarranted ostentation" and "misdirected talents." Arthur Pryor's band was deemed most inappropriate for a theatre like the Capitol; Fairbanks and his associates were unhappy about the "midnight show" presentation of *His Majesty, The American.* Columnists, city fathers, and

audiences were united in their horror over "Miss West's suggestive 'shimmy' dance and rude way of singing." (Mae had had the poor taste to dance the shimmy *sitting down*!)

The following Tuesday was a black one—not only for the Capitol troupe but for everybody in the United States. It was October 28, 1919, the day the temporary "wartime" prohibition became permanent law. Major Bowes took action and sent Arthur Pryor and his band marching. Mae West (who, years later would choose Arthur Pryor's handsome son Roger as one of her leading men in *Belle of the Nineties*) developed an intuitive case of tonsillitis after reading the notices, and did not return for the remainder of her $500 one-week contract. She was replaced by an act called "Dainty Marie." (In 1937, when workmen were doing some backstage remodeling, a jingle Mae

The grand finale of the *Demi Tasse Revue* revealed the Twinkle Star Girls on a spiral tower rotated by the hard-working Moon Boys while everybody sang "Come to the Moon," another Gershwin original, as the curtain fell.

had written on a dressing-room wall in lipstick was uncovered when a makeup shelf was removed. The verse had to do with the shimmy and the Major and was, according to *The Times*, hastily painted over.)

But ill-winded brass bands and ill-mannered shimmy dancers were only the beginning of the Capitol's troubles. There was a notice printed in the theatre's programs that read: "Every attaché of the Capitol Theatre has signed a pledge to render cheerful, willing service and to extend uniform courtesy to all patrons without expecting or accepting tips. In addition to the usual scale, each is paid an amount covering an estimate of what tips would aggregate. The management requests that patrons aid in maintaining such service and dignity and honor of the attachés by refraining from offering tips. Please do not embarrass the attachés of the theatre by proffering them tips. Your approval is sufficient reward for any service rendered."

The Capitol attachés took a dim view of this idea; in 1919 theatre tipping was a firm tradition. On the Sunday following the theatre's first hectic week of operation, they decided to show the management who was embarrassing whom when it came to tips. All but twelve of the thirty-two ushers waited until the evening's de luxe performance had begun, then changed into mufti and started picketing the auditorium, yelling at their nonstriking buddies to join them. "Their actions attracted the attention of the audience," reported *The Times*, "many of whom found the strike of more interest than the screen. The boys were ordered to leave, but they refused. Several policemen were sent from the West 47th Street Station House; they arrested Herman Ingberg, one of the strikers, upon his refusal to leave, and charged him with disorderly conduct.

"Managing Director Edward Bowes said last night that when the ushers protested he had called their attention to the fact that before accepting their jobs they had signed an agreement that they would abide by the no-tipping policy.

"The ushers said they changed their minds."

•

The patrons who had found the picketing ushers more interesting than the photoplay on the Capitol's fancy satin screen, were symptomatic. People came once to marvel at the size of the place, and stayed to yawn at the length of the show. Three hours of de luxe performance may have been worth the $2.20 a reserved seat cost—but who wanted it? In those days the average picture ran little more than sixty minutes, and two hours was as long as anyone wanted to devote to the still-faintly-disreputable business of going to the movies. Nineteen nineteen, with things still pretty much in a postwar pandemonium, was not a year of prosperity, and the Capitol's prices (plus War Tax) were more than people cared to part with. The box office went into a slump, and the operating overhead got so top-heavy that the whole enterprise was in danger of toppling down around the heads of the Messrs. Kendall, Bowes, Du Pont, and all.

One afternoon early in 1920 there strode into Messmore Kendall's twelve-room howdah atop the stumbling White Elephant of Fifty-first Street, a saturnine figure who announced himself as F. J. (Joe) Godsol. He further announced that since he had already bought Goldwyn Pictures several months before, he was now prepared to buy the Capitol—or as much of it as was for sale—to show his Goldwyn pictures in. Kendall paced his gold-leafed office while he pondered the offer. Godsol was not exactly the sort he (nor Du Pont and Bowes) would have chosen as a partner. He had made a somewhat dubious fortune during the war selling mules to the French government (who thought they were horses—possibly until one was eaten) and had been jailed as a swindler. Later he had parlayed his mule money into a second fortune in Tecla pearls "much superior to the original." Not really what you'd call a gentleman.

But after all, money talked. The last thing the patrician Kendall saw of Joe Godsol that afternoon was the spectacle of the cigar-puffing, mule-skinning, pearl-pushing multimillionaire slipping his butler a fifty-dollar bill. The deal had been made and, for better or for worse, the Capitol was saved.

Godsol's original deal with Sam Goldwyn was for the latter to stay on in an advisory capacity to the new board of directors of Goldwyn Pictures which not only included Kendall, Bowes, and T. Coleman du Pont but a strong representation from the other side of the du Pont family—investors in Godsol's Tecla Pearls and other of his enterprises. These du Ponts were then in control of the family business and didn't see eye to eye with T. Coleman, who had guided its destiny before going into politics. With the disagreeing du Ponts it was more eye *for an* eye, and the wrangles over the reorganization of Goldwyn Pictures grew so frequent and so fierce that one day Sam could stand it no longer.

"Gentlemen," he cried, "you can include me out." And with this historic utterance, Samuel Goldwyn departed, leaving behind his name (or at least the official improvement on the original Goldfish), his picture company, and a string of movie houses around the country.

The Capitol Ushers ready to trounce the Washington Senators (or the Rivoli Red Sox) in a rooftop game.

While these theatres couldn't compare in most cities with their Zukor and Fox-owned competitors, Godsol saw in them the nucleus of a larger chain that would give him exclusive houses for the showing of Goldwyn Pictures from coast to coast. Now he needed a showman to run them for him.

His eye lit on Roxy—his handsome head neither bloodied nor bowed from the debacle of the Rothapfel Unit Programme — and Godsol made him a quick proposition. Roxy gave him a quick answer, and was soon on his way to the West Coast to show Godsol what he could do with a really shaky member of the original Goldwyn chain, the California Theatre in Los Angeles. First he closed it down and called in the painters to have it done over from pit to booth in "rich gold" —already the favorite Roxy color. The orchestra pit was enlarged, the stage hung with new curtains, and Roxy started training a corps of uniformed ushers. The California reopened on November 12, 1919, with Christian Timmer, the violin virtuoso, playing Reisenfeld to the California's forty-piece orchestra.

There were the usual soloists, plus a new note in Rothapfel presentations: a name act. Will Rogers, fresh from his success in the Ziegfeld Follies of 1919 in New York, came on the stage to spin lariats and yarns and make jokes about the War Debt and Prohibition. The feature film was called *Nearly a Husband.*

Back in New York, Godsol took a long sharp look at the Capitol's books and the dismal tale they told of declining revenue, and knew that something had to be done right away. He decided that the Capitol needed Roxy's magic touch far more than the California Theatre in Los Angeles (by now well on the way to recovery), and he lost no time in recalling his staff magician to Manhattan.

This idea didn't appeal to Major Bowes one bit, but he was persuaded to go along with it when he was made a vice-president of Goldwyn Pictures at a stipend of $25,000 a year in addition to his regular salary (another $25,000) as Managing Director of the Capitol, a title which he would continue to hold.

As for Roxy, he was content, for the time being, to have Major Bowes present *him.* He had made a financial deal with Godsol that was to his increasingly hard-to-satisfy taste, and was he perfectly happy to have the Capitol programs read simply: "Presentations by S. L. Rothafel, Originator of this Form of Divertissement." The great originator had come up with another surprise that year: deciding that Rothapfel (which meant "red apple" in German) was an unpopular sort of name in those teutophobic postwar years, he simplified it to Rothafel, (which meant nothing in American). He really needn't have bothered; in a few years his nickname, Roxy, was going to be so well known everywhere that he might as well have forgotten that he had any other name.

And so it was as S. L. Rothafel that he came to the world's largest theatre and (to its owners) the world's most monstrous migraine.

On the first of June, 1920, the Capitol closed for a period of soul-searching, minor refurbishing and frantic rehearsal. On June 4 an advertisement announced: "TRIUMPHAL RE-OPENING—The Capitol—The World's Largest, Coolest, Most Beautiful Theatre. Newest, Latest, Rothafel Motion Picture-and-Music Entertainment, under the Personal Supervision of S. L. Rothafel. Edward Bowes, Managing Director." At the bottom of the ad, after the opening program had been announced, were these reassuring lines: "Prices 40c - 55c - 75c - $1 —No Higher (including War Tax)—Matinees, 30c - 40c —No Reservations! First Come First Served!"

That night Roxy gave them Erno Rapee and the Capitol Grand Orchestra playing Herbert's *American Fantasy Overture* with Singing, Tableaux, and Motion Picture Interpolation. Then a ballet, *Whispering Flowers*, by the Albertieri Dancers, followed by a Prizma Color Picture, *Hagopian, The Rug Maker.* Next, *Indian Love Lyrics* with recitation, singing and tableaux, accompanied by the orchestra. Then a newsreel, a Travelaugh, and the feature picture, a "snappy, up-to-the-minute comedy" called *Scratch My Back*, a Goldwyn Picture (of course). The whole thing, despite its surface similarity to the Capitol's original opening program of the year before, ran only two hours, and *The Times* remarked that "the Capitol's first program under Mr. Rothafel's aegis, all in all, is good entertainment and promises well for the Capitol's future."

Only one minor tragedy clouded the brilliance of the Capitol's re-opening night. Young Arthur Rothafel, Roxy's exuberant eleven-year-old, took a header down the famous white marble stairs, bloodying his nose and making extra work for the long-suffering charwomen.

The Midas of the movie palace had only to lay on his golden touch and the Capitol started to make money. Roxy had a generous budget and a staff of hand-picked experts at his command. His tasteful and fast-moving shows were soon filling the 5,300 seats for every performance which was a continuous affair now, the old reserved-seat-two-plain-and-two-de-luxe policy having been abandoned along with the $2.20 prices. While Major Bowes, USAR, sulked in stately splendor in the "front of the house" (the orchestra pit was a sort of no man's land) Major Rothafel, USMCR, was having the time of his life up on the stage.

Roxy wasn't spending all of his time in the Capitol, however. Joe Godsol had bigger and busier plans for him. Shortly after the Capitol re-opened, Godsol set about enlarging the chain of Goldwyn-controlled theatres which he now owned. He got his hands on a chain of houses in Chicago and the surrounding territory, operated by Nate Ascher. The Ascher chain had been playing second fiddle to the burgeoning Balaban & Katz empire, and Godsol, true to his old motto, "Money Talks," was prepared to give B & K a run for *their* money. With a circuit of forty theatres now in his possession, Godsol envisioned the "Capitol Wheel"—with all the theatres playing in rotation stage presentations originating at the Capitol. On Roxy's shoulders fell the burden of organizing this ambitious enterprise.

The plan called for Roxy to go to each theatre in the circuit in turn, hire orchestras and permanent soloists, choose conductors, train house managers, tutor electricians, and lay all the groundwork for the programs about to be launched. Each theatre would stage locally produced Rothafel-type presentations under Roxy's supervision until the circuit was safely established. Then, and then only, would the wheel begin to turn. At that time the show from the Capitol would move the following week to Boston and thereafter play one-week engagements at every house on the circuit. However, only the artists and scenery would be moved from theatre to theatre; the pictures would be furnished by the various Goldwyn exchanges, the music furnished by local orchestras and the scenery shifted by local stagehands who, by that time, would be thoroughly experienced in Rothafel methods.

On one of his train trips to Chicago to supervise the metamorphis of Ascher's Roosevelt Theatre, Roxy noticed that the crack *Twentieth Century Limited* was wholly lacking in external glamor as it stood waiting in the station. When he got back to New York, he suggested that a red carpet unrolled on the station platform might be just the thing to set the *Century* apart from all other trains. So delighted were New York Central officials at this idea that they gave Roxy a lifetime pass on the railroad.

With the new Capitol-Godsol circuit in mind, a smaller stage-within-a-stage was built behind the Capitol footlights; a production designed for the huge Capitol stage would be far too big for the other houses in the chain. This cramped Roxy's style somewhat, but not nearly so much as the back-breaking schedule Godsol had outlined for him. On Thursday he began rehearsals for the following week's show and continued to work on the show until it opened on Sunday afternoon. Then, after the last evening performance on Sunday, Roxy would dash to Grand Central Station and board a sleeper for some outpost like Dayton, arriving there in the morning on Monday. Rehearsals in Dayton started that day and ran on until that show opened on Wednesday; then Roxy was on the train again back to New York to keep the Capitol going.

The original plan was for Roxy to play these frantic split-weeks until all forty theatres in the chain were ready for the Capitol Wheel to spin.

But it's safe to say that he and Godsol soon found some more practical way to handle the job. Even S. L. Rothafel got tired once in a while, and New York Central officials were beginning to have misgivings about their generosity.

A Rothafel rehearsal at the Capitol was a long and noisy affair. After all the acts had been individually rehearsed in the practice halls beneath the theatre, Roxy would begin putting the whole thing together on stage in the weary hours after midnight. Sitting about fifteen rows back in the orchestra seats before a small table with a gooseneck lamp, some push buttons and a pile of scripts, blueprints and lighting charts, Roxy, in shirt sleeves and green eyeshade, would scream directions toward the stage. The auditorium's acoustics were phenomenally good, but they didn't work in reverse. The only way Roxy, with all his training on the parade grounds at Dry Tortugas, could make his threats and cajolings heard behind the footlights was to yell through a megaphone.

Then one night science caught up with art at the Capitol. Engineers of the Bell Telephone Company were experimenting on public-address systems at the time, and they were anxious to see how their new device would work in an auditorium as large as the Capitol's. Roxy permitted them to come in and set up their apparatus, and at one rehearsal they persuaded him to give up his megaphone and save his energy by using their microphone-amplifier-loudspeaker system. He was so delighted with the *vox Dei* results that he gave the experimenters from Western Electric Laboratories (Bell's development branch) permission to use an offstage dressing room as a permanent place to house their testing equipment. A few weeks later they installed microphones around the Capitol's proscenium to pick up the music of the Capitol Grand Orchestra; this they relayed by wire back to the laboratory down on West Street on the waterfront.

One of the most fascinating activities of the American Telephone and Telegraph Company those days was the operation of an infant radio station called WEAF. By the fall of 1922 WEAF was on the air every night in the week and actually had seven "sponsored" programs, though the sponsors—among them a dentifrice manufacturer, who offered a discreet talk on the care of the teeth; a political organization which gave nonpartisan "November talks" to voters; R. H. Macy, as well as Gimbel Brothers — were not allowed to mention what they made, orated for, or sold. Browning-King, Inc., the men's clothiers, sponsored an hour of rollicking melody by Anna C. Byrnes and her Orchestra, but not a word was ever said about hats, spats, or cravats.

In November of 1922 WEAF was bursting at the seams with noncommercial pride and excitement: the problem of "remote" broadcasts—those originating outside their tiny studios on Walker Street in Manhattan—had been licked. This meant that now the whole world would soon be pouring into listeners' living rooms, or at least as much of the world as could be filtered through galena crystal, cat's whisker, and earphone. On November 11, WEAF broadcast a performance of *Aida* by the Metropolitan Opera Company in the Kingsbridge Armory miles uptown (what the Met was doing in the Bronx is not disclosed). Spurred on by this success, the officials of American Tel and Tel approached Roxy and Major Bowes to see if they could broadcast an experimental program direct from the stage of the Capitol during an actual performance.

Now, for more than a year celluloid soothsayers had been predicting the immediate collapse of the motion-picture industry once people began deserting the theatres to stay home and listen to Vaughan DeLeath. The vision of a nation of stay-at-homes, glued to their sets while theatre seats gathered dust, was the same nightmare that was to haunt exhibitors again thirty years later (only with television, the nightmare came true). Major Bowes didn't like the idea at all.

"Comfort the enemy?" he snorted. "Never!"

"Applesauce," said Roxy, and started making plans to go on the air.

Major Bowes, in *robe de cuisine*, is glimpsed in his kitchen overlooking Broadway as he prepares his *spécialité*: Boston baked paper towels.

34

"HELLO, EVERYBODY . . ."

"Roxy—here is a name to conjure with! Who has not heard it—and more, who has not heard Roxy, for he is the stellar god of a new force—a "Big Timer" of the ether. When science shot lilting melodies through the air and called it radio, Roxy was born in the first pink blush of radio's morning. Before that he was S. L. Rothafel. Then came this new force. They placed a microphone before his lips, he spoke into it, and his words clutched a million hearts though the miles between were many."

AMERICAN BUSINESS RECORD,
December, 1925

The momentous event took place on Sunday evening, November 19, 1922. The audience in the theatre didn't know they were witnessing history; but WEAF's listeners — the men in lonely lighthouses, the farm families resting after chores, the city dwellers playing with their new loudspeakers —the folks in Radioland knew they were hearing something very special indeed. The program was the regular presentation of the new Capitol stage show that started that day. The highlight — "a musical event of importance"—was the playing of Richard Strauss' *Ein Heldenleben* by the specially augmented Capitol Grand Orchestra, conducted by Erno Rapee.

As Charles D. Isaacson pointed out in his introductory explanation from the stage, this was the first time the Strauss work had been presented in any theatre in America. What he failed to point out was that it was also the first time it had ever been heard along RFD #1, East Colrain, Massachusetts . . . or in the drawing room of a Murray Hill town house . . . or in a lighthouse at Barnegat Inlet. In those days of uncrowded air, a radio station with a good signal could be heard for hundreds of miles. And WEAF had a good signal.

Roxy, standing in the wings behind a microphone, described the dancers, the costumes, the scenery and the lighting. He also mentioned the name of the film playing that week on the Capitol screen, and he was sorry all the listeners couldn't see it with him because it was such a wonderful picture. He was back to a description of Mlle. Maria Gambarelli, his beloved "Gamby" (who at the moment was *en point* under a pink spotlight), before the aghast WEAF officials realized that somebody had finally given a "commercial" on their untainted air waves.

The next day the line before the box office of the Capitol was four abreast around the corner and halfway down the block. Roxy had done it again, this time proving to fight 'em you gotta join 'em.

While Major Bowes grumpily admitted that Roxy had been right, Messmore Kendall was ecstatic. He had been sold on the miracle of radio the previous evening when, at the instigation of WEAF, he had invited a group of somewhat skeptical friends to witness the broadcast from his private belvedere. That night the plate-glass window had not been lowered; but as the guests sat in what was actually the first "sponsors' booth" in the history of broadcasting, they could hear everything that was happening on stage through the horn of Kendall's Atwater Kent. The evening's moment of drama came when the glass was suddenly lowered. Sure enough: the music coming from the orchestra pit below was right on beat with the music coming out of the radio. It really worked!

Before the week was out it was decided that the Capitol shows would be a regular feature on WEAF's Sunday evening broadcasts. There was already a bundle of letters testifying to the impact of the first program; some were from amateur DX'ers who spent their evenings combing the air

The King of Radioland kept a firm grip on his scepter.

waves to log distant stations, but most were from average listeners who had happened to tune in at the right time and stayed to be enchanted. Many of them referred to "the man with the nice friendly voice who told us about the show."

Roxy canceled his trip to St. Louis, where another of Godsol's theatres was in line for his ministrations, and spent the week trying to probe the full meaning of what had happened the Sunday before. It was something very big—no one knew how big—and it was growing faster than anyone knew. And he was going to be a part of it.

By broadcast time that second Sunday Roxy could hardly wait to get to the microphone. He could see all the thousands of nice, warm-hearted people waiting to hear the "man with the nice friendly voice" again. He could see them smile as he came on. As Erno Rapee entered the pit to start the overture, Roxy got the signal from the WEAF man in the wings on the other side of the stage. He planted himself resolutely in front of the circular gadget, cleared his throat—

"Hello, everybody. This is Roxy speaking. . . ."

. . . And there he was, once a two-dollars-a-week cash boy, talking to the universe; Roxy, a Leatherneck, talking, for all he knew, to a general; Roxy, a dishwasher in an all-night greasy spoon, a book peddler, a ballplayer, a bartender, talking maybe for the President to hear!

That night he laughed as he described the antics of the pantomimists. He made little jokes of his own. He grew serious as he introduced the next piece of great music his orchestra was going to play. He burbled with joy as he followed the twirls and leaps and kicks of dainty Gamby. His voice choked with tears as he announced that Gladys Rice was going to sing a mother song.

By the time the show was over Roxy had dragged himself to such peaks of emotion, had thrown himself to such depths of pathos, had laughed and cried along with (he was certain) the folks "out there," that he found his carefully rehearsed farewell speech had left him completely. As he got the signal to sign off, he was barely able to manage the sentence that came to his lips:

"Good night . . . pleasant dreams . . . God bless you."

"God bless you" had slipped out somehow, and as he walked off the stage he decided to be more careful next time. After all, he really didn't *know* these people.

Next day Godsol burst into Roxy's office—the last person he wanted to see. "Roxy, that 'God-bless-you' stuff was great, just great—just the kind of hokum these people love. Keep it up!"

"It's out!" bellowed Roxy.

"It's in!" screamed Godsol, turning purple. But Roxy was adamant. You open your heart by accident, and Broadway tells you it's a great act. It wouldn't happen again.

On Tuesday the mail began to pile in. The week before it had been a bundle; this week it was a bushel. There were letters from Fredericksburg, Virginia; Lebanon, Pennsylvania; East Orange, New Jersey. "Your God-bless-you was a benediction," one of them said. "Your voice helped me feel that I will be blessed," wrote another. "I have found an understanding friend," penciled a third. Roxy needed to read no further.

"It's in," he said. "You bet it's in."

A contemporary ad for the RCA Radiola reflected the hold radio had gotten on the public. "Pluck the best orchestra seats right out of the air!" ran the headline. "Who would have dreamed less than two years ago that anyone owning a radio receiver could sit in the comfort and convenience of home and listen to entire productions of such operas as *Aida, Pagliacci,* and *Cavalleria Rusticana* by the San Carlo Opera Co.; the play, *The Perfect Fool* with Ed Wynn; and entire acts of *Smilin' Thru, East is West,* and *Fair and Warmer;* Roxy and his Capitol troupe, concerts by Yerkes and Paul Whiteman's Band, solos by Gadski, Lipowska and Cavalini . . . Hoffman, Hornberger and Grainger; Grantland Rice broadcasting the World Series; and the voices of the men running for office."

But not everyone owning a radio receiver sat in the comfort and convenience of home every night; Roxy saw to that. There was always a subtle plug for the movie at the Capitol, always the careless hint that while it was loads of fun to sit home and listen to the music from the theatre, it was really ever so much more fun to come to the Capitol and *see* it all happening before your own eyes.

Before many weeks had passed it became apparent that Roxy, with his "Good night . . . pleasant dreams . . . God bless you" had become famous. WEAF was forming the nucleus of a radio network—with WCAP in Washington, D.C., and WJAR in Providence, R.I.—and already the Capitol programs were being heard over half the United States. It occurred to Roxy that the idea of his standing in the wings describing what was happening on stage might soon begin to pall on the public, once the novelty of the simple fact that it was happening *at all* wore off. And so a broadcasting studio was set up in one of the rehearsal halls in the basement of the theatre, and it was here that Roxy started to preside over the first variety programs actually planned for radio. Part of the broadcast still came from the theatre upstairs—the organ, the orchestra, the magic sound of audience applause—but the rest of it was put on in the basement studio by the regular members of Roxy's troupe.

"My gang," he laughingly called them . . . and the name stuck. Roxy's Gang became as familiar to listening America as Roxy himself. There was Maria "Gamby" Gambarelli, the Capitol's prima ballerina, with her cute little songs; Douglas Stanbury, the sandy-haired baritone (they were the first "sweethearts of the air"); and there was soprano Florence Mulholland, with a voice full of heart. And "Daddy" Jim Coombs, the "sailor, beware" basso-profundo. And Beatrice Belkin. There was Eugene Ormandy, concertmaster of the Capi-

Roxy (center) and the Gang in the Capitol Theatre studio.

Roxy and the Gang visit a veterans' hospital to deliver bedside crystal sets — a favorite benevolence of Roxy's as well as a sure-fire public relations asset to the Gang.

tol Grand Orchestra, and Yascha Bunchuck, its brilliant cellist. There was Wee Willie Robyn, the half-pint baritone. And a dozen others.

By 1923 "Roxy's Gang" was a national institution. The broadcasts started at 7:20 on Sunday evening, and lasted two hours. A radio spectacular? Only in the literal sense of the word; nothing was terribly unique about a program's running two hours every week in those *dolce far niente* days before broadcasters became clock-watchers.

As radio grew to network status, so did Roxy's concept of its importance. It is interesting to speculate what might have happened if he hadn't become a national institution so early in the game; perhaps people *would* have killed the movie industry by their fascination with the new free entertainment that radio offered. Instead, Roxy became a symbol of the motion picture, and to millions he was a constant reminder to "get out and go to the movies."

Roxy, almost before he realized it, became one of the outstanding personalities of Early American Radio. His cheery greeting, "Hello, everybody," his infectious chuckle, and his general air of joshing good nature turned the shrewd showman into a sort of favorite uncle in millions of homes. The chatter of the Gang seemed as natural as any around a boardinghouse parlor. When Roxy chuckled, listeners all along the NBC Blue Network chuckled too. As one writer put it, ". . . the nation could not do without him altogether."

Roxy before the microphone was Roxy transfigured. Between his "Hello, everybody" and his "God bless you" he became confessor, confidant, cicerone, advocate of the impossible, protector of faith, and Foxy Grandpa, all in one. Roxy's Gang was the family tree from which Arthur and All the Little Godfreys sprang in later years; the lives of the Gang belonged to Roxy . . . and that meant they belonged to everyone. Gamby and Doug conducted the world's most public love affair; the very sound of her tinkling giggle and his hearty laugh would set romantic hearts pounding in a million living rooms.

All of them—Gamby, Doug, Erno, Frank, Florence, Yascha, Billy, Gene, Daddy Jim, and all the rest—were Roxy's children. And one of the features of every broadcast was the reading of letters from his "children" out there along the air lanes.

"Here is a letter that came to us this week from a little mother who lives . . . well, I'll let her tell the story. Believe me, children" (and here the voice would take on a huskiness and just the hint of a sob) "my heart is touched when I read such lovely things as this letter from this little mother. God bless her. This is what she writes: 'Dear Mr. Roxy: We live on a barge and I have two children. Well, the little girl is Margie and some time ago you said, Hello Margie well my daughter said hello she says mama he said hello. We enjoy your program every Sunday night. And we all wish you luck on your trip and all have a nice time. We love you and your sweet Gamby and we are so happy that Doug makes her so happy.' "

Then Roxy would say, "Doug, come over here a minute. Do you suppose you could sing 'Margie' right now—just a chorus, maybe? How about it, Erno? Do the boys know it? Wonderful—here we go." And there they went, bringing joy to the happiest barge on the Hudson.

"Dear Roxy: We are two little girls age nine and eleven. We are cousins and we never fail to miss your entertainment every Sunday night. Love Bertha and Flora Feeney," went another letter.

"Well, bless my soul . . . I certainly don't know any songs named Bertha or Flora . . . oh, wait . . . Daddy Jim, what was that song from *The Bertha the Nation?*" (giggles from Gamby). "So, my darling little cousins, how would it be if Daddy Jim Coombs sang 'Little Nelly Kelly, I Love You,' from Mr. George M. Cohan's new show? Just shut your eyes and pretend you are both named Nelly Kelly, and Daddy Jim will sing it just for you."

One of Roxy's favorite projects, and one, perhaps, that brought more genuine happiness to more people than anything else he ever did on the air was his campaign to provide every patient in every veterans' hospital in the nation with a little radio and earphones. Money poured in every week in response to his requests for donations to further this cause, and whenever enough had been collected to outfit another hospital, Roxy—and usually most of the Gang with him—went to the hospital to present the radios and visit with the veterans.

"It is impossible to estimate the good that Roxy has done among the battle-stricken boys of our land," editorialized the Ridgewood (N.J.) *Herald*. "Having conceived the idea of equipping every bed in every Army hospital with radio earphones, he would not rest until through the splendid efforts of his group of artists, the funds were forthcoming to accomplish that great purpose. It is unnecessary to add what this has meant to those frail, weary sufferers. If to us Roxy has become a friend, to those bedridden helpless lads he is a hero—a staunch, never-failing comrade, who has done more to ease the burden of an enforced existence than any other individual throughout the Nation.

"With the memory of those broken Khaki Lads, then, in our minds, we, too, would say, 'Good night, Roxy; God Bless You!' "

•

The almost universal love which Roxy had won throughout the length and breadth of Radioland was not shared by a few malcontents in Capitol Theatreland. Major Bowes was constantly complaining to Messmore Kendall that Roxy had turned the broadcasts into a series of weekly steps toward canonization. The programs were still bringing people to the theatre but now they were coming because it was Roxy's theatre. "Which, dammit, it is *not!*" stormed the Major.

Had the Major known what was going on in Roxy's mind—and in Roxy's office—he might not have gotten quite so riled when he switched off his radio every Sunday night.

Roxy's theatre! That was the dream. Not since Forest City had he had a theatre that was really his—all the Keith houses, the Lyric in Minneapolis, the Alhambra in Milwaukee, the Regent, the Strand, the Rialto, the Rivoli in New York, the California in Los Angeles, all the Godsol-Goldwyn houses he had commuted between so breathlessly, the Capitol . . . and none of them his. Wasn't it time, he thought, for Roxy's Theatre?

ON THIS SITE . . .

"The Roxy—and I am prouder of that name than I am of my own—will be the fulfillment of my dreams of the last fifteen years. I will be the absolute despot of it. I have always wanted to present pictures as I think they should be presented, and with the opening of the Roxy Theatre I shall be hampered in no way whatsoever in having complete control over every detail, no matter how large or small.

—S. L. ROTHAFEL, from an interview in the New York *Morning Telegraph,* December 20, 1925

It is worth noting that nearly all of New York's movie palaces were built on the ruins of a business engaged, in one way or another, in transportation. The Strand displaced the Brewster Carriage Works; the Victoria (the Rialto's predecessor at the corner of Forty-second and Broadway) was built where a blacksmith had once spread his sinewy hands; the Rivoli occupied the site of the old Palmer-Singer garage; the Capitol arose where there had been a livery stable and a filling station. And now it happened that the old car barns, on the corner of Fiftieth Street and Seventh Avenue, fell under the gaze of one Herbert Lubin, a former film producer (no kin to Sig Lubin, the Philadelphia film pioneer) who was down on his luck in the spring of 1925.

Lubin had been daydreaming about building a theatre in New York—a big one, maybe the biggest theatre in the world. Bing & Bing, the realtors who held the old car-barn property, had announced that they were willing to parcel the site; so Lubin decided that the Seventh Avenue corner (in spite of the old adage that a theatre to be successful in New York must be on Broadway) would be ideal to build his big theatre on. Lubin marched into the temporary office of the Bing & Bing agent in the old car-barn building and asked the price of the corner.

"Three million dollars ought to take it," was the reply.

"I'll give you five hundred thousand down payment on it," said Lubin (who was practically broke at the moment), and introduced himself. The agent, somewhat breathless at selling three million dollars' worth of real estate in five minutes, listened to Lubin's proposal, agreed that the theatre was a marvelous idea and further agreed to arrange time payments for Lubin on the purchase.

It took plenty of time. Lubin called on every investment house in the city. After gathering together all the cash he could from his debtors, he still was unable to raise even a down payment on the down payment. Then one day he went to see some financier friends of William Fox's; this turned the tide, and with the money put forward by them, plus stock Lubin sold to friends, contractors, seating suppliers, carpet manufacturers, and even an architect, the money was raised in time. Now Lubin was set to build his theatre, only to realize that he knew very little about the subject. He was a promoter, not a visionary—and this project needed a visionary.

He knew Roxy, just as he knew all the managers of all the theatres in New York, and he remembered that Roxy was always talking in terms of the biggest orchestra, the biggest stage, the biggest theatre—everything the biggest in the world. Now *there* was a visionary.

Lubin went to see Roxy, offered him a staggering salary, a block of stock, a percentage in the profits and—here Lubin said the magic words:

"Rothafel, we'll even name the theatre after you. We'll call it Roxy's Theatre."

Roxy became president of the company next day, and lost no time in taking leave of the Capitol Theatre to devote himself completely to the new project. It meant vanishing temporarily from public view (and hearing) and that was difficult for Roxy who had come to believe that the entire nation depended on his fatherly guidance on the radio every Sunday night. And what would become of the Gang? The final program of Roxy's Gang (. . . who could say, perhaps forever) took place on Sunday evening, July 26, 1925. The Gang had just returned from a successful personal-appearance tour in Canada, and none of them was prepared for Roxy's announcement . . . none, that is, except Major Bowes, and he wasn't exactly a part of the Gang though he had been secretly coveting membership for some time.

Roxy then explained that he was turning the Gang over to a new shepherd, Major Bowes. "Say good-by to them for me, I can't," he said to the Major. "Au revoir and God bless you," he said huskily to the folks in Radioland. And he was gone.

"Now, with a bit of a sob in our hearts, we will endeavor to make merry again," said Major Bowes, sounding unusually cheerful. Lieutenant Gitz-Rice jumped up from the piano and shouted, "Three cheers and a tiger for Roxy!" And for once, Gamby didn't giggle.

•

Work started on the Roxy Theatre right away. An opening date was announced for Christmas Day, 1926, and skeptics doubted that the builders would make it. The Roxy was built on a cost-plus basis; contractors had already learned that it was foolhardy to take on such a job for a specified sum; therefore, long before the theatre was finished, Lubin was frighteningly in debt with an overrun of more than $2,500,000. Christmas Day, 1926, passed and all Lubin found in his stocking was a yawning hole, getting bigger all the time—and the opening of the theatre was still months away. He tried desperately to raise cash, but rumors of Roxy's extravagances (the biggest pipe organ in the world, the biggest orchestra pit in the world, etc., etc.) made further investment seem doubly risky to most of the men who might ordinarily have come to Lubin's rescue.

Finally, about a week before the theatre was to open, William Fox paid a visit to the Roxy. He strolled quietly through the huge theatre, taking in everything, saying very little. Perhaps he was recalling his early beginnings . . . the shooting gallery with its Kinetoscopes . . . the City Theatre on Fourteenth Street . . . the slow growth of his chain . . . his entry into picture producing . . . his problems in finding suitable theatres to play his pictures. The cost of the Roxy—$12,000-000—represented a sum larger than his total investment in all his theatres, his studios, his film exchanges only a few years ago. Now things were different, and William Fox was riding the crest.

That night, on top of a pile of blueprints on a dusty drawing table, William Fox signed the papers that gave him controlling interest in the Roxy Theatres Corporation for $5,000,000. Lubin was off the hook—and with a profit of nearly $3,000,000 for his promotional labors. And Roxy had a rich patron who had promised to support him in the style to which he was fast growing accustomed.

One of Lubin's grandiose dreams (and one which Roxy delightedly shared) was of a chain of Roxy Theatres all around New York. There were six theatres in the original scheme, in addition to the Cathedral of the Motion Picture itself; but only one of them ever got built.

William Fox, when he entered the financial end of the Lubin-Rothafel dream world, put a crimp in the plans for the Roxy Circuit. Fox had plans of a circuit of his own . . . the great super-Fox theatres that were to spring up in Detroit, Saint Louis, Washington, San Francisco, Atlanta, and Brooklyn . . . and decided to concentrate his talents (as well as Roxy's) on these instead of the purely local New York scene. Consequently, the Roxy Mansion, on Lexington Avenue between Fifty-eighth and Fifty-ninth streets never got beyond the steam-shovel stage, and the others were forgotten . . . except for one.

Roxy's Midway Theatre, up on Broadway at Seventy-fourth Street, was the exception. Walter Ahlschlager designed it as a miniature Roxy, seating 2,657, and complete with a scale model of the Roxy Rotunda for a lobby. But the property was bought by Warner Brothers while the theatre was still under construction, and when it opened on Christmas Day, 1929, its name had become Warner's Beacon.

So Roxy had to be content with only one theatre with his name. Every time a new Fox theatre opened, he was called upon to plan the dedication ceremonies; usually they were based on the consecration of the Cathedral of the Motion picture, including the famous " . . . let there be light!" invocation and the "Star Spangled Banner" *tableau vivant.* And the "absolute despot" had a master after all.

The following pages are facsimile reproductions from *Roxy, A History* – the lavish 60-page souvenir book prepared by the editors of *The Film Daily* and presented to guests at the opening night of the Roxy.

ROXY --- A HIST
CARBARN TO
PICTURE (

November, 1926

May 13, 1926

June 26, 1926

August 31, 1926

TWO years ago antediluvian car barns occupied the valuable site on which now stands a monument to modern theatre construction — the Roxy Theatre. This transformation was achieved by utilizing every resource of the modern architect, combined with the most advanced developments

THE story of the wonders of Aladdin's lamp records no stranger magic than that wrought here by the modern hand and brain of man. Starting with the lowly site of a car barn, the four views at the right illustrate the amazing rapidity with which this old eye-sore was razed, a foundation dug and the steel skeleton erected for the new ornate structure that is destined to elicit the admiration and applause of intelligent mankind and to mark the highest point in theatre history. All this, encompassed in practically four short months, bespeaks the stirring enterprise and incredible drive of those behind the project.

ORY --- FROM A THE MOTION ATHEDRAL

known to engineering science.

The main truss, the largest ever fabricated, weighing 210 tons, forms the supporting structure which covers a plottage of over one and one-quarter acres, providing a building which can house 10,000 people under one roof.

March 11, 1927

February 5, 1927

February 1, 1927

January 14, 1927

AT the left is the second half of this epic tale in theatre building, told as graphically as only pictures can tell a story. In three months we see the naked skeleton of steel taking on the flesh and blood of concrete, marble, granite, interior decorations, exterior structural detail, massive pillars, beautifully designed friese work, ornate hangings. A magnificent structure worthy the name of the Cathedral of the Motion Picture. A blending of the finest concepts in engineering, architecture, science and art. In truth, the humble cocoon has given forth the gorgeous butterfly, resplendent in the exquisite form and color conceived by genius.

A Few Remarks on Roxy

By ROBERT E. SHERWOOD

Editor of "Life" and author of "The Road to Rome"

The author of these "few remarks" needs no introduction to any one who ever reads a popular magazine, or for that matter a so-called high-brow one. It is altogether fitting that "Bob" Sherwood, a distinguished critic of the arts, respected alike by the critical cognoscenti and the lay lover of things beautiful, gives this introduction to Roxy, with perhaps a few sidelights that are not generally known.

SEVERAL years ago, the editor of the Motion Picture News decided that it was about time for someone to select the twelve individuals who had contributed most extensively to the advancement of the silent drama as an industry and as an art.

A committee of one hundred people was named to make the final selections, and I happened (for some reason) to be a member of that committee. One of the first names on the list of twelve that I submitted was that of Samuel L. Rothafel, then commander-in-chief at the Capitol Theatre.

When the editors of the News had counted the votes, and announced the results, I discovered that a considerable majority of the committee of one hundred felt as I did. Samuel L. Rothafel was elected to the immortal dozen, and the flags on the Capitol marquee fluttered proudly.

This, it is well to observe, was before the radio craze had burst with its full fury upon a startled public.

Those votes were cast for Roxy, not because he could say "Hello Everybody!" in a cheery voice each Sunday evening—he wasn't even on speaking terms with a microphone at that time; I and many others voted for him because we knew that he had given the movies a new dignity, a new importance, in the eyes of the movie audience.

Roxy's orchestral settings for scenic films and news reels were always in perfect harmony with the subjects on the screen. They emphasized the beauty, drama and humor. In this connection, I shall never forget "Die Walkure" as played by the Capitol orchestra when the pictures of the air-ship "Los Angeles," as it arrived in New York from Germany, were first shown.

On the Air

Two years ago, Roxy announced that he was about to leave the Capitol and start up in business on his own hook. Substantiating this statement, a yawning chasm appeared in the block between Fiftieth and Fifty-first Streets on Seventh Avenue, and the public was advised that this vacancy would ultimately be filled with the biggest theatre of them all.

Now the excavators, the drillers, the riveters, the bricklayers, the painters, the decorators and the press agents have completed their mammoth task, and the paying guests are about to be asked in.

It is an exciting event in our town—exciting, not because of the opening of a gorgeous new film palace (that sort of thing is happening all the time) but because this elaborate structure glows with the warming personality of Samuel L. Rothafel.

As a radio broadcaster, Roxy has won the affectionate esteem of millions of people who, knowing him only through the sound of his voice, regard him sentimentally as a sort of combination of Little Eva and Santa Claus. They will be delighted to hear that Roxy is to talk to them again from his own luxurious studio.

For the others who, like myself, prefer to respect and admire Roxy for his work as an exhibitor of moving pictures, the opening of this new theatre is of considerably greater importance.

It means that Roxy is to have his own temple, in which he himself is the supreme high priest, at liberty to conduct services in his own way.

The ROMANCE of the ROXY

A Fascinating Stroll Through the Fairyland of the World's Largest and Greatest Theater

by

JACK ALICOATE

ROME can be seen in a day. It cannot be fully appreciated within a year of intensive sightseeing. The Louvre may be visited between breakfast and luncheon. Its art treasures invite weeks of compelling interest. So with the Roxy. A dream come true. The world's largest and greatest theater.

It is indeed a lasting monument to the greatest force for wholesome amusement the world has ever known—the motion picture. A shrine dedicated to the universal language of music. Its four walls encompass a veritable fairyland of novelty, comfort and conveniences. To view the innermost workings of this majestic temple of amusement is in itself an education. The finished product of the master craftsman, Samuel L. Rothafel, stands supreme in its remarkable and revolutionary achievements.

The first impression on entering the Roxy is that of agreeable surprise. You are prepared for the unusual. Your expectation is more than realized. Your eyes encounter so much that is out of the ordinary that you find it difficult to center your thoughts on any one feature.

The size of the stage and proscenium arch amaze you. No other structure in the world equals their proportions. From this point the theater radiates out fanlike, with a tremendous sweep of balcony at the widest point and a shimmering bronze dome overhead. It is vast. It is amazing in its sheer beauty of design and decoration. A symphony in color. A harmonious blending of luxurious draperies clothing an architectural masterpiece.

These are first general impressions. Although it is the largest theater in the world it possesses an unmistakable atmosphere of intimacy. That in itself is an architectural achievement. There are so many features of outstanding interest that you are undecided on which to center your attention first.

The stage has a distinct appeal. There is always a fascination in getting a glimpse "behind the scenes." The Roxy stage is set low, so that no matter where you sit in the auditorium you are looking down and not up. It is divided into four sections, two of which are on elevators. They can be raised or lowered at will, controlled by hydraulic electric apparatus. Complete sets can be built so that a change of scenery can be made in twenty seconds. The stage area is so vast that a performance of the most elaborate production can be given.

There is a huge cyclorama for diffusing sound. This great bulk of steel and plaster weighing ten tons can be lifted with the ease of a handkerchief. Here also is an immense curtain with its great double tableaux effects. There are magic draperies for the two-color effect that will take light and absorb it A complex system of traps is designed for the handling of scenes. As you stand in the center of this vast stage with these mammoth mechanisms all around and above you, it transports you back to childhood fancies. At last you have found the magician's workshop—the home of the genii—the castle of the giants.

Here modern science works greater wonders for your amazement and entertainment than were ever pictured in those fairy tales of long ago. For that

The Roxy's tuning Lyre, the most elaborate ever produced by the Deagan Co., and pitched to A-440 Universal Low.

is the mission of the Roxy. When you enter its portals you step magically from the drab world of confusion and cares into a fairy palace whose presiding genius entertains you royally with all the fine allurements that art, science and music can offer.

That is the spirit of this Cathedral of the Motion Picture. It is reflected in everything about you. Here in front of us is the huge pit in which the orchestra brings to you all the rich treasures that music affords. Over 100 specialists of their respective instruments. There are no less than four celebrated conductors to interpret for you your favorite compositions.

Let us tarry a moment at this great department of music. Here is the last word in a musical library. It is conceded to be the largest theater collection, topping even George Eastman's at Rochester. Ten thousand numbers and fifty thousand orchestrations. Almost unbeliveable. But here they are in these myriads of special cabinets covering the walls of the library. You will be interested to learn that the nucleus of this collection was provided by Victor Herbert's library which Mr. Rothafel purchased. It requires a small army of librarians, arrangers and copyists to properly handle them.

Here in the orchestra pit an unusual sight confronts you. It holds three immense organ consoles. A miracle of modern music come to pass with the aid of electricity. Can you picture it? A grand organ being played simultaneously by three men.

The Kimball organ is a masterpiece of construction. It is installed in special sound proof chambers

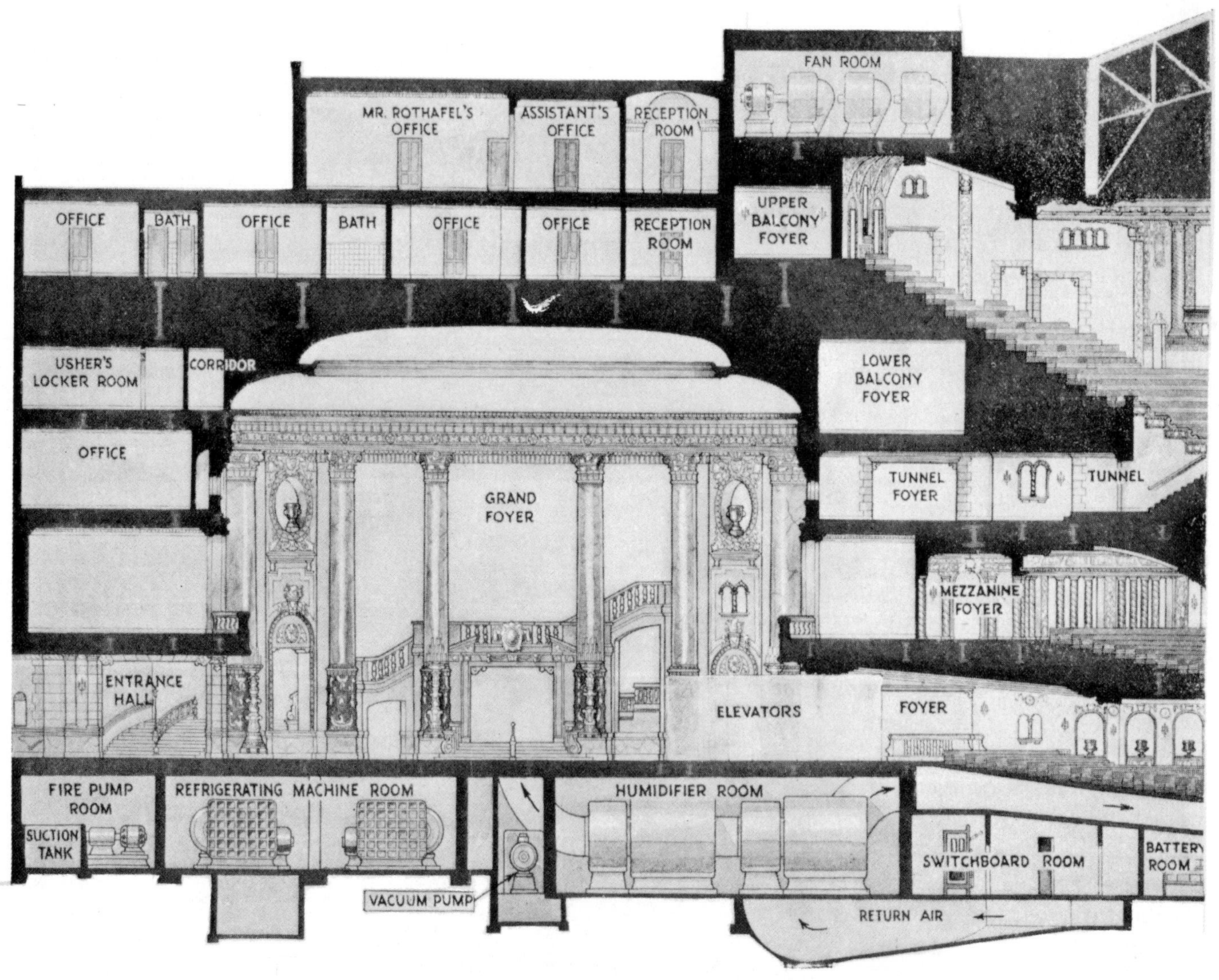

The drawing across the bottom of these two pages shows a complete cross section of the Roxy Theatre building, indicating the various innovating features. On this page is the portion of the theatre including the entrance hall, grand foyer, refrigerating

under the stage. Thus the sound comes directly from the orchestra pit. It has the properties of a symphony orchestra. It is capable of a range for musical production perhaps never before attempted on any organ.

Some conception of its magnitude can be gleaned from the fact that the main organ chamber is sixty feet long, thirteen feet deep and eighteen feet high. The tone openings into the pit are approximately sixty feet long and eight feet high, allowing the tone of the organ to reach the theater from the same position as the large symphony orchestra. This increases the possibilities of blending colors. It is ideal when playing with the orchestra. Truly it is styled "The Organ With the Million Voices." A faint conception of the electrical control of this instrument is realized when you are told that in one cable alone are over 45,000 wires.

It would seem from this that the possibilities of employing the musical Muse for your entertainment had been exhausted. But high up in the proscenium another innovation has been worked—the Deagan Chimes. They consist of 21 bells—real tower chimes, such as designed for belfry or open air use.

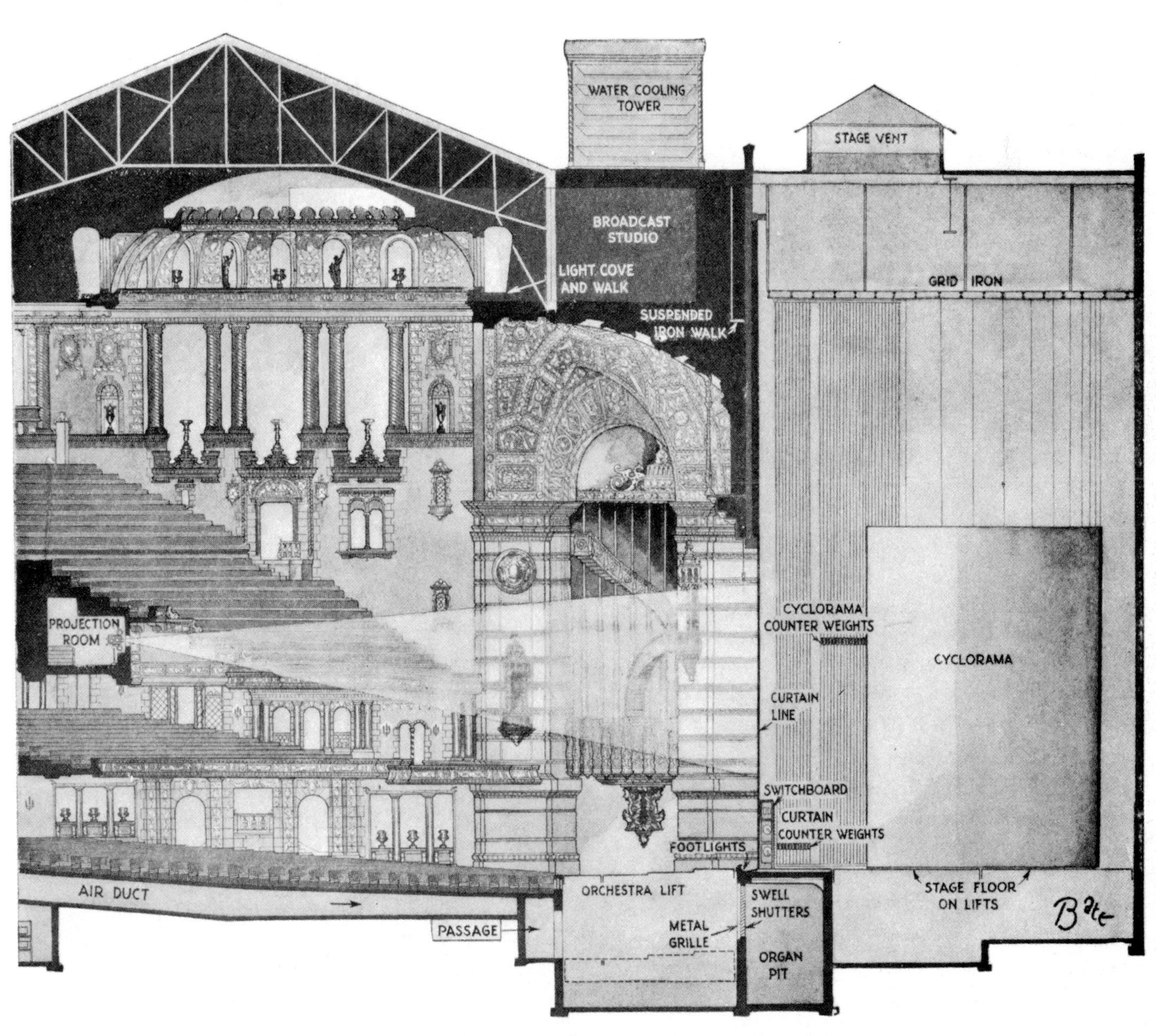

and ventilating mechanism. It will easily be seen that the auditorium incorporates many radical departures in design. The stage is located in the corner of the approximately square auditorium. Courtesy of "Science and Invention."

Do not confuse them with the ordinary chimes that you hear in the organ and played by the percussion man or drummer. So great is their vibration that it was necessary to place them behind enormous shutters to control the immense volume of sound.

Finally, the assembling of musical units is completed through a Fanfare Chamber high up in the other chambers where the organ is usually located. This fanfare is equivalent to twenty-four trumpets and sixteen trombones. Picture this galaxy of musical forces supplemented by a chorus of 100 trained voices. The possibilities can readily be imagined.

Innovations De Luxe

We are only scanning the highlights of all the rich treasures so painstakingly assembled under the roof of the Roxy for your diversion. Already you begin to realize that revolutionary things have taken place in the world of the theater. There are many more delightful innovations, too.

Without proper acoustics all this tremendous effort at musical entertainment would go for naught. So cunningly has the architectural scheme been built that the acoustics are perfect. The facilities for lowering the stage, the height of the proscenium, the flattening of the domes, the materials used, and the general contours—all these factors contribute a remarkable effect. Anyone speaking from the stage is heard in the remotest seat in the theater. We must not overlook that cut in the balcony from where the projection operates. It permits the sound to travel underneath and through into the farthest corner of the mezzanine. Seated in the balcony you realize another innovation. Because of this cut you can readily observe the people in the orchestra. Here is one of the secrets of the air of intimacy.

The placing of the projection booth in the front of the balcony you will realize is another radical departure. In most motion picture theaters the "throw" is so lengthy that the figures on the screen are often distorted. The "throw' in the Roxy, instead of being the usual two hundred and fifty feet, is exactly one hundred feet.

Projection Without Peer

The projection booth is an institution in itself. Here are sixteen operators working on various shifts —an unheard of thing in any picture theater. There are a battery of projection machines of various types such as was never before assembled in one booth. The Vitaphone with its synchronization of sound and pictures opens up an entirely new world of musical possibilities. It brings to you as you are seated in the Roxy all the greatest operatic voices as well as the work of celebrated musicians and entertainers. The unique Spoor Natural Vision invention with its third dimension achievement will have its world premiere installation in the Roxy. Another achievement that enriches the possibilities of screen entertainment. You could spend hours examining the innovations in this projection booth and find the experience vastly diverting.

Ornate detail, with an imaginative flow of rhythm and structure marks the various units of decoration.

The screen itself has not been overlooked in Mr. Rothafel's sincere desire to afford you something superlative in motion picture presentation. The screen you will note is the exact size of the picture. Instead of the usual black masking, a delicate gauze behind which there is a cyclorama of silver cloth with various colored lamps playing on it gives a soft, diffused light to the picture. A secret process of the master of modern entertainment. The result of nine years of experimentation.

It seems almost incredible that all these innovations could be prepared for your enjoyment by any one individual and offered under one roof. If any might question that Roxy is a master entertainer, here before us is the proof.

You must observe it all from the vantage point of the balcony to secure a proper perspective.

An elevator takes us comfortably and speedily to three different balcony levels—the top, middle or front sections.

Ornamental head in frieze work in the rotunda

From the balcony we gain a finer conception of the architectural design and beauty of the entire structure. An accomplishment in both engineering and architecture. But it required a harmonious decorative scheme as a proper setting. Indeed the correct decorative treatment was the final touch to properly clothe it and bring out all the beauty and brilliance of this Cathedral of the Motion Picture. The rough white plaster models were transformed into a harmonious scheme of color which with the rich velour draperies of golden brown complete the Spanish note of the architectural conception. Then the ancient spears and halberds in their artistic brackets placed about the walls—a little detail completing the picture.

Lightings

Like all works of art, proper lighting is vitally essential to bring out its intrinsic beauty. From the wrought iron fixtures the light glows without being obtrusive. The domes are illuminated so that the glow is delicately transfused to the entire auditorium. As the footlights and other lights begin to play on the stage and in the proscenium, they are reflected in this huge auditorium, creating a myriad colors and moods that are indescribable.

Such is the setting for the presentation of the new Roxy standard of motion picture entertainment. We have simply strolled through the empty theater. As Roxy's guest during the performance you realize that your comfort has not been overlooked. You sink into a richly upholstered seat. It affords unusual pleasure to realize you are not cramped for space. People pass before you to their seats without touching your knees. A new system of seat-indicator lights has guided you without confusion. Every vacant seat is instantly shown on an indicator board near the entrances. You are now ready to enjoy a few hours of entertainment, and in an atmosphere of luxury and beauty that only Roxy could have achieved. An elaborate air conditioning system purifies the very air you breathe and keeps the temperature always normal.

The muses of song feature part of the frieze design on the walls of the Roxy.

After the performance you might desire to inspect the rest of this marvelous structure. We have only viewed the theatre proper. There are the foyers, the rotunda, the halls. The administrative offices are upstairs. Here also is the broadcasting room. A fully equipped hospital. Shower baths, library, innumerable offices and rooms for eevery conceivable requirement. This achievement of one man's personality is indeed remarkable.

Other Features

It is simply additional evidence of not only foresight but the imagination that seems to be without boundary in the creation which will stand as a monument to its progenitor.

What is more, this description of the various ramifications, innovations, incidentals and essentials of the Roxy as hereabove set forth can at best be only a mere superficial rendering of an almost spiritual thing which has to be communicated in material terms. As in all efforts of this kind, it is only the tangible value on the surface that is conveyed, for the mere use of words must fail to encompass the infinite shades of meaning and romance which have been fabricated into the structure of the whole.

It is almost a pity that the public, who though appreciative of the energy and genius which has gone into this undertaking, could not have been personally present while the structure was still naked in its absence of ornate walls, finished contours and soft blendings of wall colors and decorations.

In this wise, just as a dissected anatomy proves on close inspection to a new student, to be a vastly illuminating example of the miracle of the human mechanism, so would a glimpse behind the beautiful walls and artistically ornamented pillars and draperies reveal the real significance of this gargantuan institution, cloaked and dressed over miles of cables, ducts and channels, myriads of unseen mechanical arteries and fibres, which furnish life, breath and soul to the body.

Napoleon once asked when requested to take on a new general, "What has he done?" The thought suggests The Romance of Roxy. He has developed more big theaters than any showman in the world. He is directly responsible for the modern type of presentation. He has created through sheer personality the greatest radio following in America. Last but not least, he has built the largest and greatest theater in the world—the Roxy.

Roxy himself can't get over the wonder of the Kimball Organ, which has the properties of a symphony orchestra and a main chamber 60 feet long, 18 feet high and 13 feet deep.

A HOUSE BUILT ON MERIT

Skill of hand and mind have been combined with the finest of materials to make this edifice worthy of the work of masters. On such a foundation, it cannot help but endure. May it also prosper.

By MAURICE KANN

FROM the noisy staccato of steel rivets to the soothing strains of the grand overture in less than eleven months.

That is the history written around the erection of the Roxy by the construction forces which surmounted all manner of obstacle in their fight against time. To be exact, it required ten months, 13 days to complete this wonder theater. This unparalleled record in theater building, achieved by the Chanin Construction Co., was accomplished without the slightest deviation from the high standards of quality set at the outset. Every little detail looking to the beauty of the house and comfort of patrons was catered to with the same thoroughness accorded the predominant phases of the structure.

The dream of Roxy would have failed of accomplishment had he not surrounded himself with a battery of artisans second to none in their chosen metier. With them Roxy must share the credit that goes with the consummation of such a gigantic project. The Cathedral of the Motion Picture will long perpetuate the deed of those who gave long hours of research and labor to make the Roxy a beacon of shadow-land.

Extending its invitation to lovers of the ultimate in entertainment, Roxy's exterior lighting effect has created much comment among Times Square theatergoers. The novel and striking effects are the creations of the Norden Sign Co., and Rainbow Light Inc.

Service is the watchword of the Roxy and is strikingly exemplified by the provision made for protecting patrons against the discomfiture of waiting in line for tickets. Engineers of the Automatic Ticket Register Co., have solved the problem through installation of six Gold Seal Automatic Ticket Registers, which handle the crowds swiftly and easily.

One of the unique and truly showmanship features of the Roxy is its spacious lobby. Its beautiful decorative scheme, created by A. Battisti & Son, makes the all-important first impression of the patron a lasting one. The gorgeous carpets, as well as the carpeting throughout the theater, including the celebrated "two ton" chenille, were devised and furnished by Stern Bros.

A work of impressive beauty is the grand foyer, whose beautiful decorations and draperies emphasize the element of simplicity and subdued tones throughout the house. Long conferences with the Rambusch Decorating Co., resulted in the special designs and colorings. The auditorium looks like a huge hammered bronze bowl with its deep rich plush and simple but dignified hangings with little gold and red fringe. The novel draping, the work of the Louis Kuhn Studios, is a revelation.

The lighting system was so devised that patrons at all times save when the house is darkened for effect, will be able to read their programs.

Light fixtures throughout the theater convey a

wrought iron and pitch-lamp effect. They are the conception of Robert Phillips. Glowing without being obtrusive, they evenly illuminate the domes, transfusing the glow to the whole auditorium. Frank L. Decker was in charge of the lighting effects, installed by the Hub Electric Co., which reflect in the huge auditorium a myriad of colors and moods which beggar description.

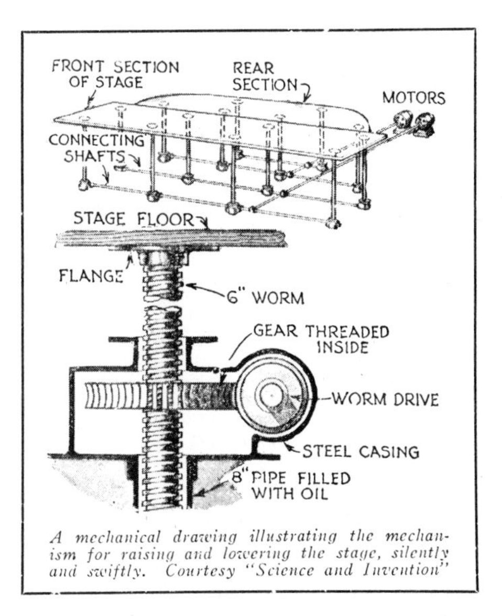

A mechanical drawing illustrating the mechanism for raising and lowering the stage, silently and swiftly. Courtesy "Science and Invention"

The stage, equipped to handle productions of great size, is divided into four sections. The first and fourth are stationary. The two center sections may be raised or lowered at will, through the arrangement perfected by Peter Clark, Inc., which firm is responsible for the revolutionary stage equipment. Hydraulic electric apparatus, furnished by the Gurney Elevator Co., enable the stage force to make a change of scenery in 20 seconds. The combination cyclorama, sounding board and sky is the first ever installed. This has been done before in plaster only but never as a combination sounding board. Worthy of special note is the manner in which the stage is dropped so that no matter where the patron sits, he is looking down and not up.

Back stage is the great ventilating system built by the Carrier Engineering Corp., the largest ever installed in a theater. It draws air from the roof, washes it, then pulls it down from the roof and through the mushrooms in the seats, where it is again washed, cleaned and re-circulated. On the hottest day of the year, the theater can be cooled to a temperature of 58 degrees.

There are three consoles in the orchestra pit. This is the first time this has been attempted; a grand organ being played by three organists at three separate consoles. The Kimball organ, especially designed for the Roxy, is the largest in the world, and one of three separate instruments in the theater, designed for three different uses. The others are a Kimball Solo Reproducing Player, installed in the rotunda, to entertain members of the Roxy Family while entering and leaving the theater, and the Kimball Broadcasting Organ, installed in the chambers opening into the broadcasting room.

In addition to the giant Kimball in the orchestra pit, up in the proscenium is a set of 21 grand chimes manufactured by J. C. Deagan, Inc. The chimes have individual electric action and damper for each tone and really are designed for belfry or open air use where great volume is necessary. They are played from the organ console.

The unique location of the projection room—in a cut in the balcony—has a three-fold purpose, the bettering of the theater's acoustics, the improvement of projection and creation of an atmosphere of intimacy despite the theater's size. The distance from the booth to the screen—the "throw" of the picture is exactly 100 feet, instead of the customary 250 feet. All distortion is eliminated by this innovation. Sixteen projectionists are on the various shifts which will be entrusted with the projection of pictures for Roxy's gang. Three different types of projectors demonstrate that the Roxy is keeping pace with developments so that the ultimate in motion picture entertainment may at all times be provided for patrons of the Roxy. There is the standard Simplex projector, the product of the International Projector Corp., a special

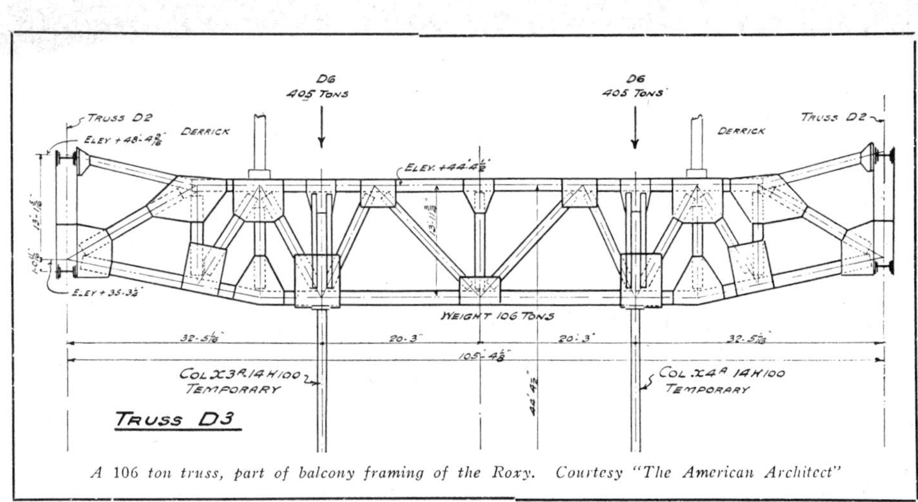

A 106 ton truss, part of balcony framing of the Roxy. Courtesy "The American Architect"

projector for Vitaphone presentations, and one for proiection of stereoscopic pictures, first of which now is being produced in Hollywood by the Spoor Natural Vision Process. Six high intensity arc lamps, installed by Hall & Connolly, are used in the booth, four for the Simplex projector and two for use with the Natural Vision Pictures. Five others are used with the Brenkert floodlight. Hoffman & Soons supplied the rheostats for the exacting requirements. Equipment experts agree that the system used at the Roxy, which was designed by the National Theater Equipment Corp., represents tremendous and revolutionary strides in projection.

Comfort of patrons, a predominant factor in the equipping of the Roxy, attains a new degree of emphasis in the 6,200 seats of the Roxy. This important feature was entrusted to the Heywood-Wakefield Co., and the result strikes a new note of convenience and comfort. There is plenty of room between rows, so patrons may walk between rows without touching a person seated. Every seat is raised so that patrons look down on the stage, thus eliminating neck-craning in order to obtain a complete view. Patrons are guided to their seats by indicator lights, part of the efficient Bilmarjac Seat Indicator System. This keeps ushers advised of empty seats through an intricate mechanism which assures safety and efficiency.

No need to detail the size of the chimes; the picture speaks for itself. They weigh 10,000 pounds.

Bilmarjac is a new seat indicator system and its selection for the Roxy is a tribute to its efficiency, in view of the short time it has been on the market. It consists primarily of aisle boxes three inches wide and nine inches long placed on the side of aisle seats. These aisle boxes have a perpendicular row of circular discs corresponding to the number of seats in that particular aisle. When a seat is vacated, the disc corresponding to that seat automatically lights. Under the last disc on each aisle box is a lighted square indicator that bears the designation of the particular row of seats. On the side of the aisle box facing the stage is another light that burns continuously.

It throws an amber light four inches wide by fourteen inches long, covering the space between the rows of seats. This assists patrons in entering and leaving the aisle, shows whether there is a step up or down or whether the aisle floor is level with the seat floor. The system is controlled by a large indicator board on the wall near the entrances to the auditorium. This board holds light discs corresponding to every seat, which light immediately when a seat is vacated. The head usher thus knows to which aisles to direct patrons and the aisle usher is guided by the aisle box indicator lights.

Fourteen Steinway pianos are used throughout the house, in keeping with the high music standards set for the Roxy.

While out of sight of the audience, the elaborate equipment for lighting plays an importance in the conduct of the Roxy which cannot be overestimated. The giant switchboard, designed especially for the Cathedral of the Motion Picture by the Decker Co., has a capacity three times as great as any theater today in the world.

Construction of this intricate mechanism was entrusted to the Hub Electric Co., and its installation to H. Alexander, Inc. The electrical consumption and equipment of the Roxy is sufficient to light a town of 25,000 population. A feature is the ground glass, which by means of periscoping reflectors enables the switchboard operator to look at the front of the stage with the lighting and performers working on it. Seven rooms are devoted to transformers, switchboards and equipment.

Revolutionary ideas in electrical engineering characterize the lighting system for stage and auditorium. The length of the switchboard is approximately twenty-two feet, declared to set a new record for compact construction.

Another Roxy achievement, invisible to the audience, is the plumbing system, which honeycombs the entire structure. V. S. Rittenhouse, Inc., is responsible for the many innovations and efficient equipment which marks this phase of Roxy service. Wash room facilities for 10,000 people are provided. The piping system approximates three and one-half miles of pipe, laid under unusual and difficult conditions peculiar to the theater's construction. The water supply tank of the theatre is on the roof, 150 feet above the stage, and from this point the entire house is served. Some conception of the elaborate nature of the system may be gained from the fact that one run of piping extends in twisted fashion 700 feet.

Many handicaps were overcome by E. G. Woolfolk & Co., in the installation of the direct heating system. An additional main was supplied by the New York Steam Co. to supply the necessary steam.

To Percival R. Moses, consulting engineer on heating, cooling, ventilating and electrical plumbing, credit is due for the finished result which fully meets the Roxy requirements A. O. Greist was supervising engineer on post at the theatre.

Two hundred and fifty tons of steel were used. The Levering & Garrigues Co. furnished the steel for the structure, with H. G. Balcom and Samuel Klein consulting engineers on steel and steel design respectively.

Ornamental drinking fountains, placed in convenient locations, were supplied by the San-Dri Co., and are in keeping with the beauty of the theatre's surroundings.

There are many features for the safety of Roxy patrons, among which is a complete hospital, equipped to handle any emergency. There are male and female wards, presided over by a representative staff of physicians and surgeons. While, necessarily, the hospital is a miniature in size, it is complete in every respect.

Close-up of one of the ornamental figures adorning balcony arches.

There also is complete equipment for nose and throat treatment, a special clinic to check threatened colds. While these facilities are designed primarily for employes, they also are available to patrons. Two nurses are in attendance.

Facilities for comfort are everywhere. Compact telephone booths are artistically located in convenient corners where members of the Roxy Family are assured private conversation without waiting in line. A broad stairway leads to the rest and smoking rooms.

A musical library, estimated as the largest theatre collection, has been installed at the Roxy. It contains 10,000 selections and 50,000 orchestrations, the nucleus provided by Victor Herbert's library which Roxy recently purchased. Seventy-five especially constructed asbestos lined cabinets with automatic locks already line the walls of the library which is guarded by three librarians with Abraham Gaber in charge.

Rotundas or vestibules at the exits, add to the comfort of patrons, for they protect them from the elements in event of inclement weather. The foyer and rotunda will accommodate 4,000 persons, which, with the seating capacity of over 6,000 enables the Roxy to accommodate 10,000 persons.

Executive offices have been carefully planned, for they are to serve as the base of operations of the Roxy organization. They are designed to take care of executive needs as the Roxy circuit expands. Roxy's office contains a wide variety of instruments necessary to his complete supervision of the theatre's operation. There are direct and inter-communicating telephones, a large radio, a Duo-Art piano, phonograph, little colored ground glass bulbs for signaling and various other devices which will mark a new high standard of direct contact with every phase of the theater's activity.

Not far removed from Roxy's study are special rooms for various assistants. All are arranged and furnished to assure the maximum of efficiency and comfort.

On the top floor is the broadcasting room, where Roxy's Gang will perform for the vast radio following they command. Its equipment reflects the preparations Roxy has made to enable him to continue as a prime favorite among radio listeners.

Wrecking for the Roxy started Nov. 22, 1925, with excavation work begun Dec. 19. Work on the foundation started March 22, 1926, with steel construction begun April 28.

The magnitude of the Roxy Theatre is emphasized by the fact that the following amounts of material were used in construction: 250 tons of steel, 4,000,000 bricks, 1,100 tons of art plaster and lime, 30,000 yards of metal lath, 40,000 yards of burlap, reinforcing plaster imported from Dundee, Scotland; 500,000 feet of galvanized tie wire, 700,000 feet of channel iron, used in holding plaster in place; 200,000 feet of one and one-half inch angle iron, 70 tons of modeling clay.

That is the story of the Roxy, a history marked by devotion to an ideal, the combining of brains and brawn to make the theatre one of the show places of the world—a shrine for lovers of entertainment—in every sense The Cathedral of the Motion Picture.

THE ARCHITECT'S VIVID CONCEPTION—NOW A REALITY

The Aim for Beauty and Utility

The Problems Confronting the Creators of the Roxy

Designed by

W. W. AHLSCHLAGER

Architect

LET us study the problems and the objectives which confronted those who were identified with the Roxy at the time of its inception. The theatre's enormous seating capacity necessitated an auditorium beyond precedent in any theatre in existence. Its large size occasioned problems of visibility, acoustics, and what is today called "traffic conditions,"—again without precedent. Its great size occasioned most careful study to bring about a feeling of intimacy, rather than a feeling of largeness and openness, that occurs in our stadiums of today, and in our gardens (athletic). Its great size occasioned untried problems in interior design and decorating.

The Roxy seats over 6,000, to which must be added the standee space for 500, and further appointments for an additional 2500 patrons standing in line formation in various foyers, after having purchased tickets. This makes a grand total of 9,272 patrons who paid admissions, all of whom are under the roof of the Roxy at one time. To these figures must be added the orchestra, approximating 100 musicians; the ballet and chorus, approximating the same number; and the organization and management corps of 300 more.

This makes a grand total of

almost 10,000 people within the theatre at one time.

Six stories with individual elevator, back stage dressing rooms, rehearsal rooms, library and broadcasting department, together with a stage of enormous proportions, the rear half of which is electrically elevated and lowered—together with an electrically elevated orchestra pit, and console pits, and three organs—provide the vehicle for the Roxy presentation.

A score of sub-departments almost too numerous to mention, which come under the department of service, are provided throughout the building, such as rest rooms, wash rooms and smoking rooms, library, hospital rooms, ushers' wash-rooms and locker rooms, and drill rooms, and two floors of executive offices the latter again reached by private elevators.

Above and at left, examples of the warmth in architectural values that feature the parts of the structure that will be seen in perspective. It is a distinct departure from the conventional Oriental. Photos by Tebbs and Knell, Inc.

Many unique and previously untried methods of design have been installed in the Roxy, for the better presentation of its programs, outstanding among which probably is the location of the projecting room in the cut out portion of the balcony.

Below, a close-up of the detail showing the judicious merging of the modern Italian and Spanish feeling in old gold filigree, varied only in high lights and shadows by the lighting.

Architecturally and decoratively, the primary thought has been to use a treatment which should give the feeling of intimacy, which heretofore has been found only in the small legitimate theatres of today; and decoratively, the choice of design and color has been principally induced by a maximum use of restraint and negation. Modern and recent large theatre design seemingly has been based principally upon the theory of outdoing one's neighbor in the use of marbles, ornamental plaster, and polychromed pigments, and the trend of interpretation of such houses has been that if a new house has outdone a previous house, at least in quantitative measurement of these aforementioned objectives in design, then it shall be considered as being a more successful design than that of its neighbor.

"AN ACRE OF SEATS IN A GARDEN OF DREAMS"

"In our big modern movie palaces there are collected the most gorgeous rugs, furniture and fixtures that money can produce. No kings or emperors have wandered through more luxurious surroundings. In a sense, these theatres are social safety valves in that the public can partake of the same luxuries as the rich and use them to the same full extent."

—HAROLD W. RAMBUSCH
in AMERICAN THEATRES OF TODAY, 1929

The United States in the Twenties was dotted with a thousand Xanadus. Decreed by some local (or chain-owning) Kubla Khan, these pleasure domes gave expression to the most secret and polychrome dreams of a whole group of architects who might otherwise have gone through life doomed to turning out churches, hotels, banks and high schools. The architecture of the movie palace was a triumph of suppressed desire and its practitioners ranged in style from the purely classic to a wildly abandoned eclectic that could only have come from men who, like the Khan himself, "on honeydew had fed, and drunk the milk of Paradise."

Things were not always easy for these misunderstood sybarites of the drawing board, and most of their straight-laced McKim, Meade & White-washed colleagues looked down on them. They were fair game for critics and academicians who rent their garments each time an Angkor Wat appeared on Main Street or a no-so-Petit Trianon reared its balustrade over the corner of Elm and

The Paradise Theatre, Chicago. Designed by John Eberson, 1928; sculpture by Lorado Taft. Paradise lost, it was demolished in 1956.

Broad. "This irresponsible reproduction of all the great architectural treasures of the ages," wrote one horrified savant, "so cheapens public taste that one wonders if a whole generation is not now arising whose artistic appreciation will be so warped that in years to come, Americans visiting the great sites of antiquity will be heard to remark: 'So this is the Taj Mahal; pshaw . . . the Oriental Theatre at home is twice as big and has electric lights besides.' "

There is no doubt that movie-palace architecture had a strong effect on public taste but not necessarily the one the critics had foreseen. World fairs, the works of Richard Halliburton, the travelogues of Burton Holmes, and the movies themselves had given people an appetite for the exotic, and they clamored for more of the same on private terms. Whole colonies of Spanish haciendas sprang up outside city limits, their pastel stucco walls rising undaunted out of Minnesota snowdrifts, their crinkly tile roofs bravely gathering the rains of Rhode Island in April. There were villages of two-family Tudor cottages, each with artfully set patches of tapestry brick speckling their oh-so-antique half-timber fronts, and each with a thatched well-house sheltering a Willys-Knight. And there were suburban casbahs chock-a-block with three-bedroom mosques, radio aerials stretching from minaret to minaret.

The tantalizing dream world of the movie palace came true over and over again as shrewd real estate operators cashed in on the public's thirst for fantasy. And many a starry-eyed housewife vowed she would be happy for the rest of her days serving Cream of Wheat in a breakfast nook with wrought-iron gates and imitation-tile Congoleum on the floor.

Disillusionment came fast as the stucco cracked, the thatch mildewed and the minarets tilted. But instead of shaking the public's faith in the architecture of make-believe, it only whetted tastes for bigger and gaudier and more stately movie mansions. In them people found escape from the ugliness of the cities, and from the crumbling boredom of life in the jerry-built Alcazar Gardens, Stratford Manors and Mecca Heights. For the movie palace architect was an escape artist. It was his mission to build new dream worlds for the disillusioned; and as he piled detail on detail, each prism, each gilded cherub, every jewel-eyed dragon became part of a whole . . . a feast for the eye, a catapult for the imagination.

For Atmospheric Theatres

The space requirements of **Brenograph Junior** as shown above for concealed work are as follows: 24″ wide, 30″ high and 30″ deep.

A canopy of clouds moving across a field of twinkling stars is used in the atmospheric theatre to complete the illusion that the pictures are being viewed beneath nocturnal skies.

Back stage, also, scenic-moving effects of all kinds can be projected from overhead, or from the wings, with a suitable back drop. In the small theatre, the back stage effects can be controlled from the projection room.

Brenograph Junior with its motor driven effects and automatic operation is ideal for these purposes.

When equipped with a short pedestal as shown on the left, **Brenograph Junior** occupies but small space. It may easily be concealed in a recess or compartment so as to hide the origin of the animated scenic effects from the theatre's patrons, and further increase their mystification and strengthen their illusion.

Brenograph Junior is especially adapted to short range work for covering large areas. One, two or four units are used depending on shape and size of area to be covered.

Cat. No. FJ-100—Brenkert Brenograph, Jr. complete as above illustrated with lamphousing and stand, effect holder, universal electric motor with variable speed control with fleecy-cloud effect complete in casing, 10 ft. connecting cord with plug and switch, any focus projection lens from 3″ to 12″. (No mazda lamp)..$290.00

Eberson's first "atmospheric," the Majestic in Houston, had ceiling fans under the balcony to stir up the atmosphere in non-air-conditioned 1922.

Cloudland created

There were two major schools of movie palace design: the *standard* (or "hard-top"), which had its precedent in the opera house and vaudeville theatre but which grew more exotic as the decade progressed . . . and the *atmospheric* (or "stars-and-clouds"), which borrowed from Nature and the more flamboyant landscape gardeners of the past.

Two individuals stand out among all the hundreds of architects who practiced the arts of illusion during the golden age of the movie palace. Not only were they the most prolific, but their places in the history of the art are undisputed. One was Thomas W. Lamb, dean of the *standard* school and the first major architect to make his name in movie theatres; the other was John Eberson, creator of the *atmospheric,* whose influence on the climate of moviegoing in the Twenties was both original and enchanting.

Eberson's first architectural job came along shortly after he had come to the United States and settled in St. Louis. It was not a theatre at all but a porch for a lady named Mrs. Sheehan of Hamilton, Ohio—a three-sided Ionic affair tacked on to her Victorian dwelling. This was in 1908 and his commission on the job was $20. In a few years he turned to theatre design, working with a promoter of small-town "opera houses." Together they traveled through the Midwest to find likely towns where the promoter would sell the citizens on the need for a theatre and Eberson would design it. They were so successful that young Eberson soon came to be known as "Opera House John." In 1922, after he had graduated to the designing of theatres for big-time vaudeville and movies, Eberson pulled his ace card: Holblitzelle's Majestic Theatre in Houston, Texas — the world's first atmospheric theatre.

In an Eberson atmospheric house, the auditorium was (to quote him) "a magnificent amphitheatre under a glorious moonlit sky . . . an Italian garden, a Persian court, a Spanish patio, or a mystic Egyptian temple-yard . . . where friendly stars twinkled and wisps of cloud drifted." Eberson was archeologist, weather man and landscape gardener rolled into one, and the combination made wonderful box office.

He had many imitators and some of the best-known theatre architects, who had created styles and huge reputations of their own, were called upon to design an atmospheric once in a while. But if some of these copies were larger or fancier or cloudier, none had quite the same air of midsummer's night in dreamland as the Eberson originals.

John Eberson was born in Austria with a European's appreciation of the great American need for illusion, and in composing the scenario for a new atmospheric, he proved he could write as colorfully as he could design. This is how he described a new project in 1926:

I am working on a French interpretation of an atmospheric theatre—the Garden of the Tuileries. We picture a Louis sending a message through the land calling for painters, sculptors, gardeners, artisans of all kinds. And he gives the command to transform the spacious lawns lying in front of his palace into a festive ground, as he is going to entertain his grandees and dames at a glorious magic night feast.

Months of artful effort and vast energy are devoted to the transformation and for the festive decoration of the lawns. Gigantic arches, enchanting colonnades, illuminated lattice garden houses, mystic pyrotechnic effects all silhouetted against the entrancing moonlit sky of a beautiful

This Andalusian bonbon is the Tampa (Florida) Theatre, complete with doves circling the balcony, peacocks preening on the organ grilles, and Christopher Columbus (below) discovering the orchestra pit.

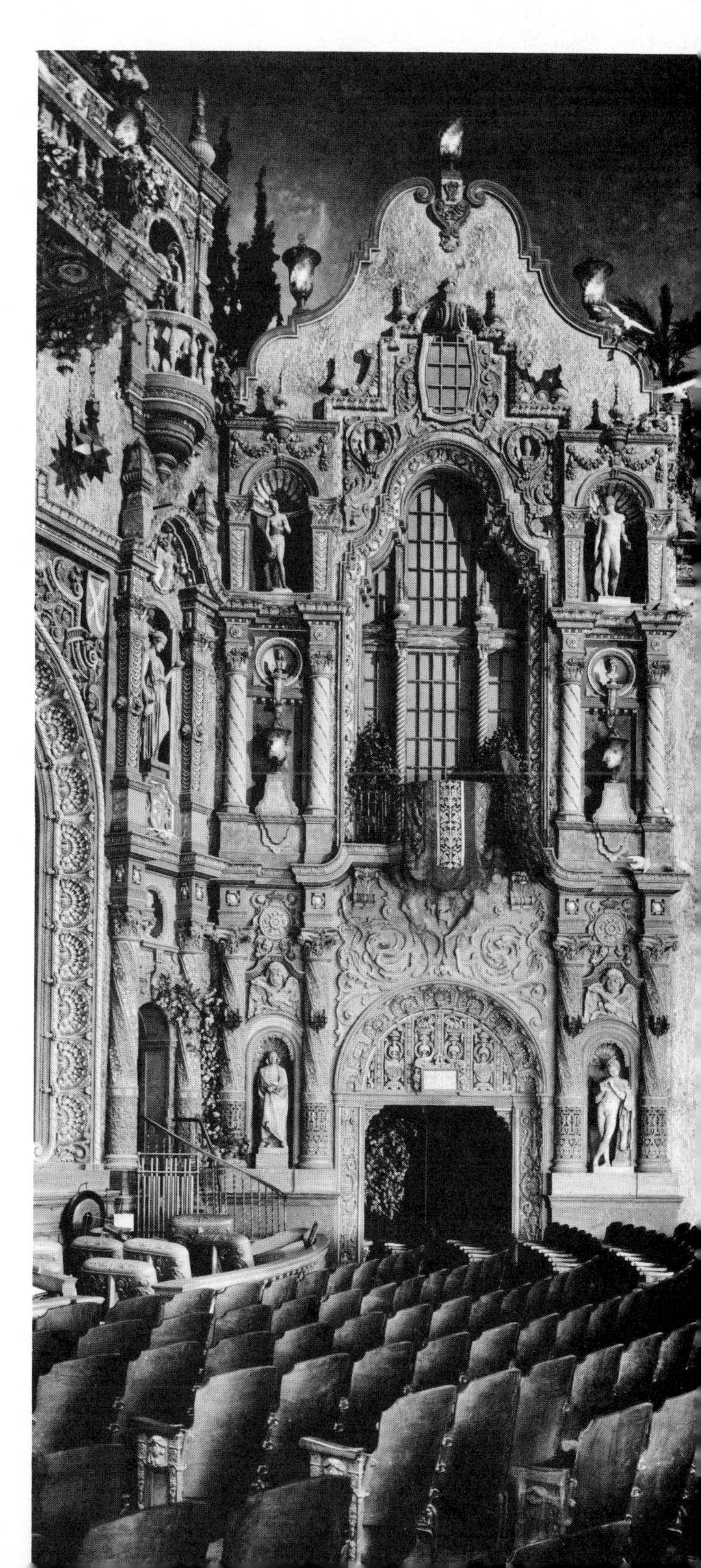

summer night. Surprises, illuminated fountains, music niches, lovers' lanes—a marvelous setting for a fantastic artful dance, the frills of the satin and silk-gowned nobles, the coquettish silk and ruffle-covered damsels, the air laden with the sweet perfume of jasmine.

Theatres were opening like morning glories all over the place in those days, and Eberson was bored with the sameness of their stuffy, academic interiors. Movie-palace architecture, an art form less than ten years old, already seemed to be in a red-plush rut as designers kept hauling in more crystal chandeliers, more marble balustrades and more dizzying domes every time a new foundation was dug. So Eberson set about to bring a breath of fresh air into the whole overstuffed concept of theatre design.

His plan for the Majestic literally blew the roof off all the old ideas; so far as Houstonians of that

more innocent day could tell, the Majestic had no roof at all. They found themselves sitting in an open Italian garden, its travertine walls topped by pergolas, classic temples and—shades of Mrs. Sheehan — twin Porches of the Maidens. Vines grew all over everything and stuffed peacocks paused to be admired atop the organ grilles; the proscenium arch, done like a huge gateway, had a real tile roof. And in the plaster firmament overhead there twinkled a constellation of electric stars laced by lazily floating magic lantern clouds.

One of Eberson's later achievements was the Avalon Theatre in Chicago—a Persian palace out on Stoney Island Avenue that would have made Xerxes tear his turban in envy. "It is here," rhapsodized the poet-architect, "that the royal nabobs and lords gather to barter and exchange everything from fruit to human souls. Behold the quietness and shade of this small pool where the numerous birds of brilliant plumage fly for refuge from the midday sun, to plunge unmolested in the clear cool water. Behold, the water carriers pause for a short siesta in the heat of the day to gaze in the pool of the Bridal Fountain and dream, perhaps—of more mysterious moonlight. The beholding of such scenes inspired the work which led to the creation of the sweet and mysterious interior of the Avalon Theatre."

Shortly after the Avalon opened, a hasty phone call from the manager indicated that while things were undeniably mysterious in the theatre, they were far from sweet. Would Mr. Eberson please come out right away? When he arrived, the manager led him to a balcony seat overlooking the auditorium and asked Eberson to sit there for a while, watching what was going on below.

What he saw was puzzling: on the right side of the theatre, patrons were relaxed in proper Saracenic contemplation of the silver screen. But on the other side there was a continual getting up,

When Loew's State opened in Louisville in 1928, the boys in Eberson's drafting room planned a surprise for the boss: the coffered lobby ceiling featured busts of Beethoven, Dante, Socrates and dozens of other notables including Eberson (top) himself.

The auditorium of the Avalon in Chicago, showing a portion of the lavish permanent inner stage set, the imported Persian mosaics and — in its niche — the Bridal Fountain.

going up the aisle and returning. What could be causing this hegira on the left?

When he went downstairs to observe the fidgety half of the audience at close range, Eberson found the answer. Nestled in a mosaic niche beneath the minarets housing the organ pipes was the Bridal Fountain. Its appeal had not ceased with the birds of brilliant plumage or the dreamy water carriers; so irresistible was its tinkle that anyone sitting within earshot was soon overcome by the power of suggestion. A minor adjustment to the plumbing of the pool, and peace was restored to the Courtyard of the Shahs.

John Eberson, always true to his slogan, "Prepare Practical Plans for Pretty Playhouses—Please Patrons—Pay Profits," designed nearly a hundred atmospherics before the end of the Golden Age. Regardless of the lilting alliteration of his "nine little P's," to call these awe-inspiring theatres "pretty playhouses" was to do them a major disservice. But they did please patrons, and they *did* pay profits.

For, with all their Persian-carpeted flights of fancy, they cost about one-fourth as much to build as the standard crystal-and-damask-marble-and-ormolu models. The simple plaster vault of the ceiling, with its projected clouds and handful of low-wattage stars, was economical in comparison with classic domes, ornamental beams, stupendous chandeliers. Most of Eberson's decorative details—gazebos, trellises, columns, arches, cherubim and seraphim—were made of cast-plaster. And many stock models of these items popped up time after time in Eberson atmospherics around the country, differing only in the peacocks they supported or the amount of arbutus that trailed around them. (All were turned out by a concern that specialized in outfitting theatres called Michelangelo Studios, John Eberson, Proprietor). Since atmospherics were comparatively cheap to build and simple to maintain, their popularity with theatre chain

Avalon patrons under the Bridal Fountain's spell found solace in the Harem Parlor (upper left) or the Caliph's Den (upper right). Tots were tended in a nursery (below) with Lindy soloing on the ceiling; no one knows where the little door at the rear led, or if it came with a bottle marked "Drink Me."

owners was not so much a matter of esthetics as of hard cash.

A few of the outstanding Eberson theatres around the United States were Loew's Valencia and Paradise in New York; the Avalon, the Paradise, the Capitol in Chicago; the Tampa in Tampa, and the Olympia in Miami; the Paramount (née Riviera) in Omaha, the Grand Riviera and Riviera Annex in Detroit, the Orpheum (pre-atmospheric) in Wichita, the State in Kalamazoo and the Capitol in Flint, Michigan; the Capitol in Grand Island, Nebraska; the Uptown in Kansas City, the Paramount in Nashville, Tennessee; the Majestic theatres in Houston and San Antonio; Loew's theatres in Akron, Louisville and Richmond, Virginia; and the John P. Harris Memorial Theatre in McKeesport, Pa.

The Paradise Theatre in Faribault, Minnesota, was not an Eberson house. It was small—850 seats—but it *was* a movie palace, and it was unquestionably atmospheric. Posted by the switchboard was this notice that established, in language that was simple yet worthy of a passage from Genesis, the little theatre's kinship with the greater Paradises in Chicago and New York:

PLEASE DO NOT TURN ON THE CLOUDS
UNTIL THE SHOW STARTS.
BE SURE THE STARS ARE TURNED OFF
WHEN LEAVING.

The Paradise in The Bronx was Eberson at his Italo-Flamboyant best. Patient caryatids flank Lorenzo de' Medici (above) as he watches the stage show. The proscenium (upper right) displays the asbestos curtain, while overhead float the clouds of a misty Brenograph dawn. A local zoning regulation on the Grand Concourse (the Rue de la Paix of the Bronx) forbade projecting marquees and towering vertical signs, hence the restrained façade (right).

LOEW'S PARADISE THEATRE
MARION DAVIES IN "MARIANNE"
DAVE SCHOOLER & HIS SERENADERS
IN "CAFE DE PAREE" FROM THE
CAPITOL THEATRE WITH E. BOREO
LOEWS
CONTINUOUS FROM 11 A.M. MARION DAVIES IN "MARIANNE" A SCREEN EPIC
DAVE SCHOOLER IN "CAFE DE PAREE" WITH EMIL BOREO - A LAVISH SHOW

Hoffman & Henon's design for the Stanley Theatre in Philadelphia was pure Adam up to the footlights; onstage the permanent set for the screen was a Market Street version of Scheherazade's boudoir.

Where the Mind is free to frolic

Thomas W. Lamb may have lacked some of John Eberson's meteorological magic, but he was one of the busiest and best known of all the movie palace architects. Lamb's career spanned the whole panorama of movie house design, from the prehistory of the nickelodeon days, through the Golden Age, and into the blue-mirror-and-chrome-stair-rail era of the decline and fall.

He was born in Dundee, Scotland, in 1887 and came to the United States when he was twelve. After graduating in architecture from Cooper Union in New York, he went to work for the city as a building inspector, marking time until he could win his first architectural assignment. It came in 1909 when he was engaged by William Fox to design the City Theatre on Fourteenth Street, then New York's movie midway. During his lifetime, Lamb designed more than three hundred theatres in all parts of the world, and had several notable achievements in nontheatrical architecture as well. Madison Square Garden is one of them; another, though it was never built, was his design for the Palace of the Soviets in Moscow. He received honorable mention in the international competition held in 1932 for his design of a panic-proof Palace with enough exits to permit twenty thousand Russians to get out in a hurry.

When William Fox approached the young architect on the matter of planning the City Theatre, he specified that it was to be built according to "high class" ideas, and in 1909 this meant two balconies and an orchestra pit. It was Fox's first real theatre as well as Lamb's, and Lamb had little else to guide him other than the fundamentals learned at Cooper Union and a remark from Fox to the effect that he was to allow space for a projection booth.

Three years and several theatres later, Lamb was retained by H. N. Marvin to design the Regent Theatre, the remarkable house in lower Harlem that was to be Roxy's first steppingstone in New York. Lamb gave the Regent a façade modeled on pure Italian Renaissance lines (see page 34) with an arcade of store fronts reminiscent of the Palazzo del Consiglio in Verona and two well-proportioned *loggie* looking out over Seventh Avenue. Although the Regent had been conceived from the outset as a movie house, Lamb still designed it as a proper theatre, with a roomy stage flanked by opera boxes. The Regent's sight lines were exemplary, however, as Lamb put in a single balcony with a gentle slope instead of the usual double-cliff-hangers that marked most theatres of 1913.

Supporting columns for the balcony were set behind the last row of orchestra seats, thus avoiding another hazard of theatregoing: the peekaboo chair. The Regent was a success architecturally, and though Roxy made some structural changes when he arrived (the removal of the projection booth to the orchestra floor and the construction of an office where it had been) the theatre on 116th Street made a name for young Thomas Lamb as an architect of unusual ability.

When Moe and Mitchell Mark saw the Regent, Lamb was promptly retained to design their new theatre at Forty-seventh Street and Broadway: the Strand. For the Strand Lamb went classical and planned an auditorium with Corinthian columns on either side of the proscenium and a huge pancaked Wedgwood bowl of a dome over the balcony. The Strand, opened in 1914, was followed by another Lamb success, the Rialto, two years later. Here the style was pure 1916 Adam. The stageless Rialto emphasized Roxy's belief at the time that photoplays and music were the thing, and that variety acts should stick with B. F. Keith. The Rialto incorporated a number of other new ideas, notably in lighting, as practical electrical dimmers had been developed that allowed Roxy to fade one color into another by means of lights concealed in the coves of the dome and around the orchestra pit — the famous Rialto "color harmonies."

Lamb's next theatre was the Rivoli (1917), northward again on Broadway. The auditorium was a pleasant example of his by-now-familiar Adam style, with the arched organ grilles, the dome and the bas-relief sounding board over the orchestra platform. The Rivoli was stageless, and was designed to follow the Rialto's orchestra-soloists-and-pictures policy. But Lamb's most notable achievement at the Rivoli was in recreating the Parthenon of Athens, in all its Doric sublimity, out of white glazed terra cotta for the façade.

Broadway was enchanted, and so was Messmore Kendall who was shopping around for an architect. In 1918, Lamb was chosen to design the Capitol for Kendall (see page 57) and from that time on his fortunes were made. For many years afterward he hewed closely to the brothers Adam for inspiration, so much so that one critic suggested, only half in jest, that the firm's name be changed to "Lamb & Adam."

Lamb defended his unswerving allegiance to the Adam brothers on the grounds that he felt this style of decoration reflected the mood and preference of the American people. His Fox theatre in Philadelphia, Albee theatres in Cincinnati and Brooklyn, Strand in Brooklyn, Keith's 86th Street, Loew's State and Academy of Music in New York—in addition to the Rialto, Rivoli and Capitol—are all prime examples of Lamb's Adam Period.

About 1925 he began to sense a change in public tastes. "I noted a lessening in the response of the average patron to the charm of architectural backgrounds patterned after the works of the Adam brothers. There was an underlying demand for something more gay, more flashy—a development for which there is much precedent in the history of architecture. For this reason I began to favor in my design an entirely different style, leaning

This early rendering for the San Francisco Fox was Lamb's bow to the French baroque; details were changed in actual construction to include a more restrained wall treatment and a Mighty Wurlitzer instead of odd upright instrument in orchestra pit.

toward the periods of Louis XVI and the very rich productions in the Italian Baroque style."

Lamb's sumptuous Loew's Midland Theatre in Kansas City was a monument to this change of heart. It was Louis XVI with a vengeance, and was the first theatre Lamb designed of what he called the "de luxe" type. "De luxe" to Lamb was almost synonymous to the 1909 "high class" style he was called upon to use in his first movie house, William Fox's City Theatre; both had two balconies, the only difference being that in a de luxe house the first balcony consisted of a small section of seats tucked under the main balcony and usually referred to in movie-palace parlance as the "loges." Lamb disapproved of loges on the ground that they made for poor sight lines, formed acoustical pockets, and were unnecessarily costly to build. But theatre owners loved them because they were able to charge extra prices for loge seats—and moviegoers loved them because the higher admission price provided them with a bit of class distinction. To say nothing of wonderful places to neck during the feature.

Lamb designed a number of theatres in the French manner, including Warner's Hollywood on Broadway in New York, a large and luxurious house that opened too late—after the advent of talkies—to be a financial success. After floundering on a straight movie policy during the Thirties and early Forties, it became a legitimate house, finally to find success as the Mark Hellinger, home of *My Fair Lady* for the last eon.

The movie palace architect's most valuable ally was the decorator, for without him the pleasure domes would have been as barren as dirigible hangars. All through the planning and building stages they worked together to create just the right

The Picture Gallery in San Francisco Fox mezzanine promenade was Mrs. Fox's pride. She selected its works of art on a European shopping spree of Marco Polo proportions, was later criticized for wild extravagance.

The Fox Theatre in Detroit was one of two identical theatres designed by C. Howard Crane; its twin is the Fox in St. Louis. The style might be described as Siamese Byzantine, or perhaps William Fox was closer when he called it "the Eve Leo style" in tribute to Mrs. Fox, who obviously had a hand in it. Whatever the style, it was a movie palace in every sense; its Wurlitzer was a mate to the one in the New York Paramount, and there was a smaller Möller organ in the lobby; its oval auditorium could seat 5,042 people in colonnaded splendor; the globe of its immense chandelier was 13 feet in diameter and was made up of 1,244 pieces of jeweled glass; its every niche, cornice and alcove was peopled with deities, basilisks, chimeras, dragons, butterflies and ring-nosed lions.

effect of awe mingled with euphoria on the absorbent ids of moviegoers. The decorator usually moved in after the structure was fairly well completed to deck the hall with boughs of gold leaf and all the other trappings that were his stock in trade.

Though there were a number of first-class professional theatre decorators in the field, Mrs. William Fox was not one of them. From days of watching the cash box and nights of poring over plans while her husband built his theatre chain from a few ten-cent shows to what finally became —for a few months—the giant of all the chains, Mrs. Fox had developed the business acumen of Hetty Green combined with the decorative flair of a demented Elsie DeWolfe. She climbed stairs to factories and lofts to see with her own eyes how furniture and decorations were made. She knew all the problems: even seat upholstery had to be tested for the effects of perspiration, of Lucky Tiger hair pomade, and for wearing qualities versus cost. And the saving of her husband's money was one of her passions.

Dealers would try to buy her with subtle lures which were, to Eve Leo Fox, about as subtle as an under-the-table pinch on the knee. When a new contract was being considered they would show her some tempting item in stock—a tapestry, a carved chair or an oil painting—which would be offered her as a gift. According to Upton Sinclair, her husband's biographer, Mrs. Fox would say: "How much does it cost?" The answer might be $500 or $1,500.

"All right. If you can afford to make me a present like that, you can afford to deduct that amount from the Fox Company's bill. Please do so." Then, her deal having been made, Mrs. Fox would leave and await delivery. What could you do with a woman like that?

When the Fox theatres in St. Louis and Detroit (identical twins designed by C. Howard Crane — see page 110) were being built, Mrs. Fox commuted between New York and the other two cities to carry on her work of patrolling suppliers. Two firms had originally made bids for decorating the theatres and had gotten into a quarrel. William Fox didn't fancy the job of deciding between them; it was, as he said, "a question of the Rolls-Royce style of theatre decoration as opposed to the Hispano-Suiza style." Rather than make a choice, he decided that both theatres would be decorated in the "Eve Leo style." She did so well and saved him so much money that she forthwith became official decorator for the Fox chain.

Thomas Lamb's great San Francisco Fox was her masterpiece. This time she forsook factories and lofts and went to Europe instead, returning with a boatload of treasures that had San Franciscans goggle-eyed on opening night (see page 107). But the first pinch of the Depression was just being felt, and Mrs. Fox was loudly criticized by some of her husband's lieutenants for her extravagance. If they had seen her haggling like a rug merchant in Continental curio shops, they might have been less caustic. And in the eyes of at least one young gentleman of San Francisco, Mrs. Fox was completely vindicated; he spent an entire afternoon—while his parents were in the auditorium being dazzled by the Fanchon & Marco spectacle and the Mightiest Wurlitzer west of the Rockies—sitting in all the thrones in the lobby, one after another.

•

The Louis XVI and Italian Baroque designs soon led Lamb into more and more flamboyant experiments until, near the end of the Twenties, he had thrown purism to the winds in favor of Hindu, Chinese, Persian, Spanish, and Romanesque themes. An interesting illustration of this new mood was a series of theatres designed for the Loew chain. Loew's State in Syracuse, New York, was the prototype of three Oriental extravaganzas that made his early Adam efforts look like Quaker meetinghouses. The Syracuse theatre was opened in 1929, Loew's 175th Street Theatre in New York in 1930, and Loew's 72nd Street (also in New York) in 1932. But the casual patron would have been hard put to tell the lobbies of the three theatres apart; the Romanesque "hard-top" auditoriums of Loew's State in Syracuse and Loew's 175th Street were identical. Loew's 72nd Street was atmospheric, however, and quite different. This trio of Hindu temples represented a great economy to the Loew chain because the same castings for ornamental plaster work, the same sets of detail drawings, identical carpeting and light fixtures — even the same elephants on the newel posts — were used in all three (see page 117).

"The Grand Foyer," wrote Lamb, in describing the theatre in Syracuse, "is like a temple of gold set with colored jewels, the largest and most precious of which is a sumptuous mural. It represents a festive procession all in Oriental splendor, with elephants, horses, slaves, princes and horsemen, all silhouetted against a deep-blue night sky. It is pageantry in its most elaborate form, and immediately casts a spell of the mysterious and, to the Occidental mind, of the exceptional. Passing on into the inner foyers and the mezzanine promenade, one continues in the same Indo-Persian style with elaborate ornamentation both in relief and in painting, all conspiring to create an effect thoroughly foreign to our Western minds. These exotic ornaments, colors, and scenes are particularly effective in creating an atmosphere in which the mind is free to frolic and becomes receptive to entertainment.

"The auditorium itself is also very much permeated by the Orient but it is not pure and unadulterated like the foyers and vestibules. It is the European Byzantine Romanesque, which is the Orient as it came to us through the merchants of Venice, those great traders who brought the East and its art back to Europe in their minds, as they brought the cargoes in their ships."

Thomas Lamb, who could wax almost as rhapsodic as the Atmospheric Laureate, John Eberson himself, looked upon these Indo-Persian-European-Byzantine-Romanesque theatres as his greatest

Lamb leaned heavily on this column. It appeared first in Loew's State in Syracuse, later on 175th Street and 72nd Street in New York (see following pages) as did a number of other Hindu conceits off the same blueprints.

Loew's 72nd Street Theatre was a far cry from Lamb's early Adam-inspired work. Designed in collaboration with John Eberson (who created the auditorium), the 72nd St's huge lantern over proscenium (above) seemed to float in clouds; balcony (lower right) featured illuminated joss houses, bronze incense burners and a tented ceiling over upper section. Ceiling of lobby (upper right) was a golden grille, cove-lit above. New York's East Side mourned when theatre was razed in 1961.

successes. Certainly, as reflections of the change in public tastes during the Twenties, they were all that movie palaces possibly could be.

When Loew's asked Lamb to come up with an atmospheric theatre, he balked at first. After all, he felt, this was Eberson's province. And previous Eberson imitators, notably the Boller Brothers in the Midwest (see color pages), had not covered themselves with glory. But Loew's, convinced that Lamb could do anything, persuaded him to try.

"One of my architect friends," observed Lamb, "has been very successful in presenting to the public through the medium of the motion picture, a theatre of a type called the 'Atmospheric Theatre,' wherein sky effects are used in place of the

Lobby of Loew's 175th Street (view from promenade, left) and lobby of Loew's 72nd Street (above) had much in common, including elephantine newel posts (below). Note how carpeting and rubber matting at 72nd Street's entrance matched, pattern for pattern, where they met.

usual ornate ceilings, and the sidewalls of the auditorium indicate scenes of the interiors of patios and decorated garden walls. My personal opinion is that this type of work will not be lasting. My objection to it, mainly, is that valuable space is used up on each side of the auditorium for effects that otherwise could be used for seats. Another thing, these various effects and ornamental details are very likely to be accumulators of dust and dirt, therefore increasing greatly the cost of upkeep.

"However, the fact remains that in a community having a number of theatres of the regular type, one of these 'Atmospheric' theatres is quite wel-

come and will retain its novelty character for a considerable length of time. At present I have the good fortune of being engaged to design five of these 'Atmospheric' structures to be located in various cities and in such widely separated points as New York City and Cape Town, South Africa."

Lamb was not only a master of words, he was good at eating them as well. His atmospheric Loew's Pitkin Theatre in Brooklyn and the auditorium of his Loew's Triboro Theatre, while lacking some of the subtlety of the Eberson touch, turned out quite well, and did, indeed, "retain their novelty character for a considerable length of time."

A pot-pourri of Loew-Lamb delights: the Buddha in the balcony (72nd Street); the camel on the sorcerer's chair and the surprisingly French ladies' parlor (175th Street); the cloister of the dervishes, the great chandelier and wall of a thousand mirrors (72nd Street); and the lost baby (State, Syracuse) all made moviegoing exciting.

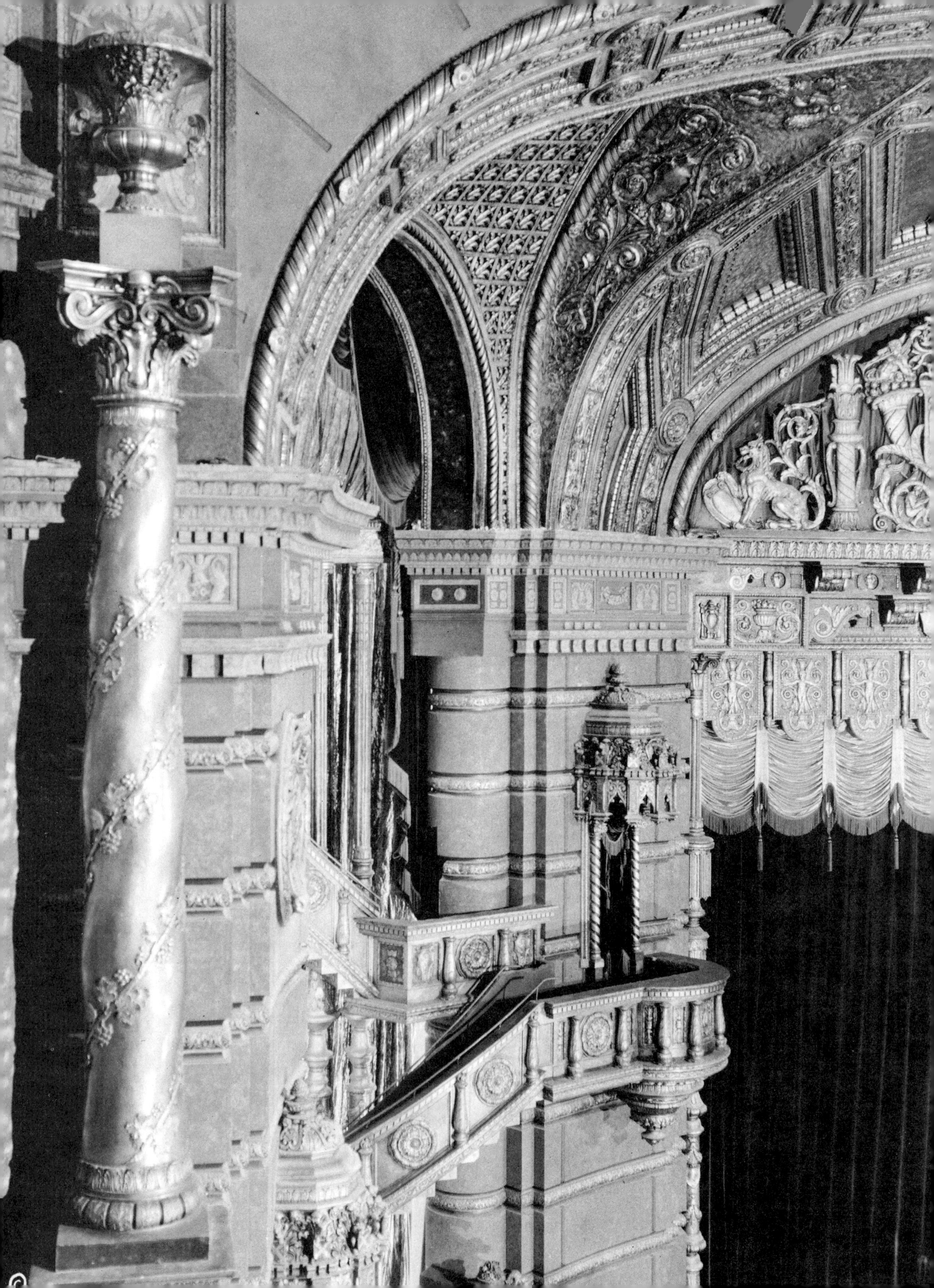

"Everything in tones of antique gold . . ."

Most architects (except those working for William Fox) dealt with professional decorating establishments like that of Harold W. Rambusch of New York. Rambusch and his associates were furbishers of wide reputation, and during the Twenties they gilded scores of architectural lilies including many theatres for Thomas Lamb, a number for Eugene De Rosa of New York, C. W. and George L. Rapp's luxurious Loew's Kings Theatre in Brooklyn, and a number of the other more plush palaces around the country. But the brightest star in the Rambusch crown was the Roxy Theatre, designed by architect Walter W. Ahlschlager of Chicago—based on an idea by S. L. Rothafel.

For the Cathedral of the Motion Picture, Ahlschlager borrowed twisted columns from the baldachino in Saint Peter's at Rome, pulpits (see drawing) from Santa Maria Novella in Florence, a plateresque *retablo* (over stage) from an unidentified Spanish church; pointed Gothic arch (above staircases) further demonstrated his catholic tastes.

The Rothafel idea was not just one idea; it was dozens of ideas every day. As an Ahlschlager aide remarked years later, "If most of his brainstorms hadn't turned out to be such good ones, I'm sure we would have walked off the job in the first three months. Roxy thought he knew it all . . . and the funny thing about it is, he did."

Among the things Roxy knew was what he wanted. When he summoned Rambusch to talk decorating, as the theatre was beginning to take shape, the interview was brief. "Harold, I see my Theatre like the Inside of a Great Bronze Bowl," he said, speaking in the Roxian Upper Case he reserved for pronouncements of special importance. "Everything in tones of Antique Gold. Warm. Very, very Rich. Gorgeous."

Pondering this dictum, Rambusch returned to his studio, and after carefully studying Ahlschlager's drawings for the interior, asked his associate, Leif Neandross, to make a rendering of the auditorium to submit to Roxy (see frontispiece). Their reward was a grunt, a sigh, and a silent bear hug. Soon Rambusch artisans were transforming the raw plaster interior of the theatre into the biggest bronze bowl in Christendom.

Of course, the "bronze-bowl" effect applied only to the burnished metallic color scheme that Roxy had specified. Walter Ahlschlager's breathtaking structure was far more than a mere bronze bowl; it was the climax of twenty years of evolution in one of the richest and most imaginative and transitory schools of architecture since the discovery of the keystone. It was, indeed, the Cathedral of the Motion Picture—the high-water mark of the golden age of the movie palace.

The motif for the Roxy Theatre, inside and out, was drawn from the plateresque school—an exuberant grafting of Renaissance details on Gothic forms with fanciful Moorish overtones. The style

ROXY-9-

Drawing by Helen E. Hokinson;
Copr. © 1929 The New Yorker Magazine, Inc.

"Mama — does God live here?"

When you entered the Roxy Rotunda you knew you were *somewhere.* Twelve huge green marble columns supported the dome from which hung a 20-foot chandelier; on the marble floor lay "the largest oval rug in the world," weighing 2½ tons. Twenty-five hundred patient patrons (including the well-Sunday-schooled little girl in the cartoon above) could wait there for pews in the Cathedral of the Motion Picture.

took its name from the ancient Spanish *plateros*—silversmiths whose designs were marked by minute and lavish detail. Ahlschlager added touches of his own, and, like every other theatre architect, dipped freely into many an architectural source-book. But the total effect of the design, from the magnificent Fiftieth Street façade reminiscent of the Cathedral at Valladolid, to the almost overwhelming golden *retablo* surmounting the proscenium, was plateresque. It was a little bit of Salamanca on Seventh Avenue.

Rambusch heightened Ahlschlager's plateresque mood through his use of forests of gold leaf in a variety of tones ranging from palest white gold to an antique bronze the color of old church bells. He specified deep crimson hangings in the arches and niches of the auditorium and wine red for the seat coverings, but when it came to choosing carpeting for the aisles none of his designs seemed to please Roxy. Finally he was inspired by a paisley shawl he saw on somebody's baby grand; Roxy was delighted, and soon miles of paisley broadloom were rolling down the aisles.

The famous Rotunda rug, measuring fifty-eight by forty-one feet and weighing over two tons ("the largest oval rug in the world") was inspired by something else Rambusch might have seen on a baby grand—a microphone of the quaint "perforated doughnut" design. Roxy was proud of his radio fame, indeed owed much of his ultimate success to his appearances on the "Roxy's Gang" programs. Rambusch's design was just the thing and the great rug was woven by Mohawk Mills at a cost of $15,000. In its center was a huge microphone monogrammed with four "R"s; the main body of the huge oval was more of Rambusch's paisley pattern, and around the border were symbolic loops of movie film. A special chewing-gum squad was appointed to give it its nightly cleaning after 30,000 Doublemint-careless patrons had marched over it during the day.

The Rotunda (ushers were sacked for calling it the "lobby") was one of the most overwhelming public rooms this side of Napoleon's Tomb. Twelve gigantic columns of *verde antique* marble, five stories tall and standing on golden bases carved with high-relief plateresque detail, rose from the polished honey-colored marble floor. Guests entered from the street lobby through an outer foyer—a fairly low-ceilinged chamber artfully designed to bring a "gee whiz" reaction when the Rotunda was reached. Here they stopped dead in their tracks before crossing the seventy feet to the polished bronze doors to the orchestra section or turning left to take the Grand Staircase to the balcony or—luxury of luxuries—proceeding to the broad steps leading to the famous Roxy over-

stuffed loges. The steps to these more costly and comfortable seats were flanked by two life-sized bronze statues — *Le Reveil* and *Le Sommeil* by Jules Cheret. For the improvident or the indecisive who had neglected to buy an extra-price ticket at the box office, there was a desk conveniently placed on the landing where regular admission tickets might be redeemed (at a price) for tickets to the loges.

On the opposite side of the Rotunda was a mezzanine where patrons of the loges might admire the *objets d'art* or gaze down on the general-admission rabble below. Here a huge arched window of amber stained glass, grilled over by a replica of the gates to the side chapel in the cathedral at Cuenca and set with a heraldic shield of colored glass, poured oceans of sunlight over the scene all day long.

Over the entrance to the outer foyer was a musicians' gallery where a three-manual Kimball pipe organ filled the Rotunda with subdued music. The organ could be played automatically by rolls, or manually; usually it was rolls in the daytime and live organist from six in the evening (the hour for the ushers' Changing of the Guard ceremony) onward. The balcony of the musicians' gallery was ornamented by a ram's head, classic swags, and six griffons, the whole being completely gold-leafed as was nearly every exposed surface that wasn't marble or crystal or travertine throughout the entire Rotunda.

At various levels around the curved walls of the Rotunda were little balconies and belvederes, some no wider than a bowman's slot, where different prospects of the great room could be studied. Floating above it all was a tremendous dome whose perimeter was molded with high-relief figures of charioteers and cornucopias and lyres and other things classical. Amber cove lighting heightened the intaglio effect of these figures and shed a warm glow over the area below. Hanging from the center of the dome was a spectacular twenty-foot chandelier made up of twenty-four nine-branched candelabra arranged in tiers like a wedding cake; the entire fixture was drenched in a torrent of sparkling rock-crystal prisms. It was lowered on a windlass above the dome when its bulbs needed changing or its prisms polishing, and it was one of the things that made The Roxy The Roxy.

The Rotunda and its adjoining foyers and staircases could hold 2,500 patrons at one time—all waiting patiently to gain the auditorium. With all this grandeur to feast the eye, soft carpeting to rest the feet, pipe-organ music to soothe the impatient breast, waiting was no great hardship for worshipers at beauty's throne. There were oil paintings to admire *(The Sacrifice of the Bull* by Benjamin West hung on the way to the Gentlemen's Smoking Room; *The Rape of Europa* by Vitello d'Impolito hung on the Mezzanine Promenade; and *The Assumption of the Virgin* by Luca Giordano hung on the Grand Staircase to the left). And there was always plenty of good reading in the latest issue of *The Roxy Weekly Review*—a twenty-four-page "magazine to take home"—which contained articles on everything from milady's fashions to sports to cabaret life to profiles of Roxy Gang members to blurbs on coming attractions. In addition there were carefully detailed program notes on the current divertissement, a complete listing of the entire Roxy staff, and a weekly editorial by Roxy himself. Parishioners not fortunate (or faithful) enough to visit the Cathedral of the Motion Picture every week could subscribe to the *Weekly Review* for two dollars a year.

The first person to buy a regular-priced admission to the Roxy (opening-night tickets at eleven dollars, engraved invitations, or personalized gold pencils don't count) was Herbert H. Pohl, a young German immigrant who was making his living

Two drowsy French ladies, ignoring swarms of rude cupids, guarded the staircase to the extra-price loges.

selling a device called the Hush-O-Phone. The theatre's purchasing agent, William E. Atkinson, had promised to give Pohl an order and had asked him to come to see him on March 12, 1927, the day the Roxy opened its doors to the general public. Pohl arrived at the theatre bright and early but not before several thousand first-dayers who were lined up along Fiftieth Street waiting for the box office to open, and he had to elbow his way to the executive entrance to keep his appointment. He had promised himself that if the sale were a good one, he would blow part of his commission on a wiener schnitzel at Luchow's; Atkinson came through with an even bigger order than Pohl had hoped for, and he decided to knock off for the day in celebration—and to start by seeing the first show at the Roxy, since he was already within its very walls.

He hated the thought of going outside and standing at the end of what was by then a line clear back to Sixth Avenue, and, glancing at his watch, he discovered that it was only a few minutes before the doors were to open. "From my vantage point inside the theatre," he recalls, "having found a way from the offices to the lobby, I watched the drillmaster barking his commands at his usher platoon and marching them off in all sorts of directions including, of course, in the direction of the entrance doors to bring the inva-

The gilded catafalque (upper left) is a detail from the Rotunda musicians' gallery which housed a pipe organ. The great amber glass window (lower left) let in sunlight all day. The crystal chandelier (above) was lowered by windlass for polishing and rebulbing. Note portion of the chariot-race frieze around dome.

sion from the outside to a proper halt. I simply marched with them right to the cashier's window in the outer lobby just as they opened the doors to let the crowd come in, and by use of this strategy and through no malice aforethought, obtained ticket No. 1. Realizing its value for generations to come, I requested the ticket-taker please not to tear it, and, having won my point victoriously, put it into the breast pocket of my coat over my heart!"

Pohl's recollection of that first show is a montage of rising orchestra pits and organ consoles, Gamby and her Floral Ballet, curtains opening and closing, movies going on and off, color, music and excitement. But the theatre itself remains a clearer memory. "I looked at numerous paintings hanging along the theatre walls; no one would have dared at that time to show modern art or abstracts or, heaven forbid, a Picasso in those plush surroundings. An air of great elegance prevailed throughout that theatre and whoever talked dared not to do so but in whispers. Compared to the Paramount Theatre which featured the popular Paul Whiteman as well as such momentary attractions as the boop-boop-a-doop Kane girl, Roxy had to be super-colossal in order to attract customers to come to see his shows, and they came from everywhere, partly, no doubt, to see the beauty of his theatre."

Shortly after the opening performance, Pohl wrote a note to Roxy and enclosed his precious "ticket No. 1" with his compliments. He received a reply that read, "I am very grateful to you for sending back the first ticket. We shall frame it

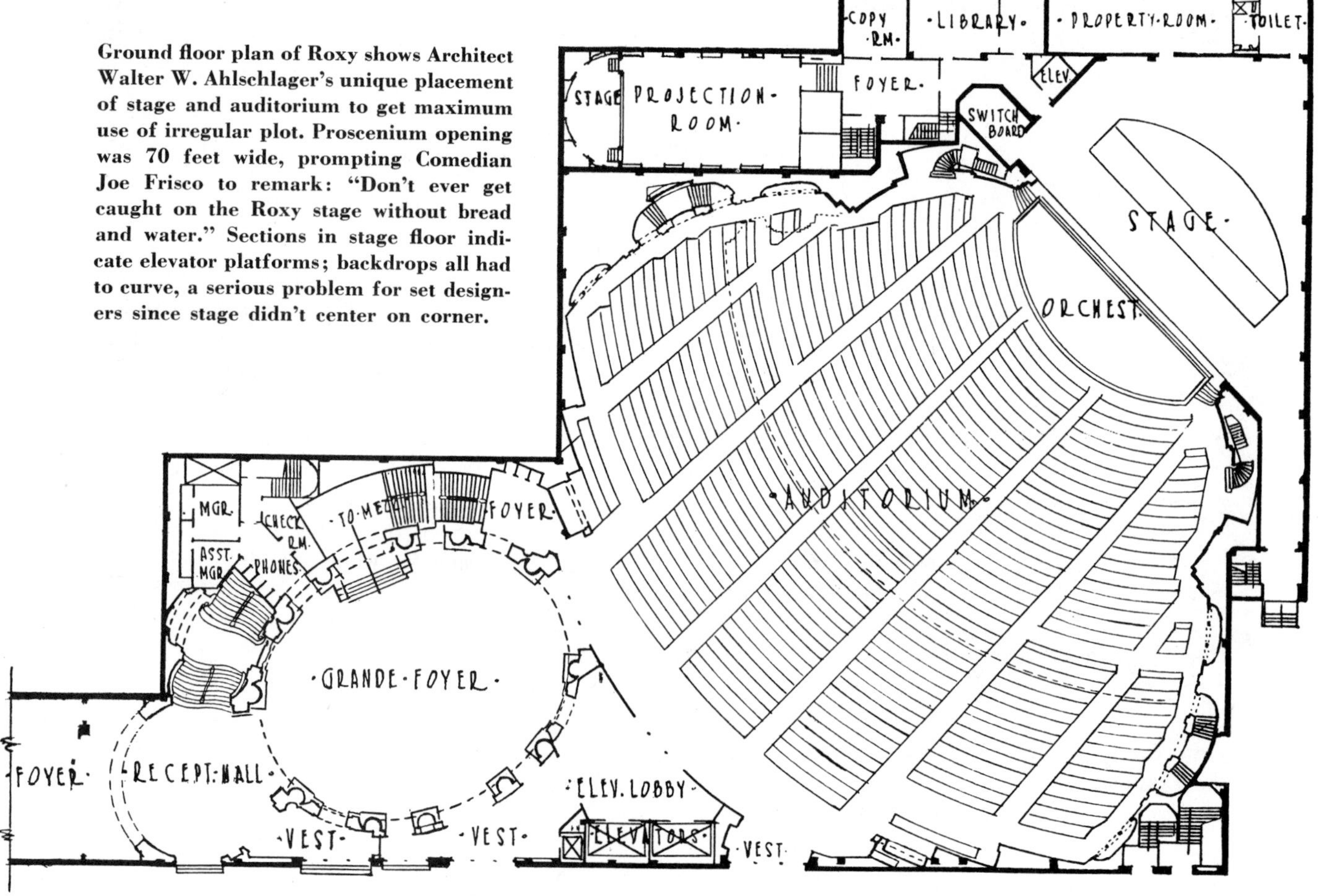

Ground floor plan of Roxy shows Architect Walter W. Ahlschlager's unique placement of stage and auditorium to get maximum use of irregular plot. Proscenium opening was 70 feet wide, prompting Comedian Joe Frisco to remark: "Don't ever get caught on the Roxy stage without bread and water." Sections in stage floor indicate elevator platforms; backdrops all had to curve, a serious problem for set designers since stage didn't center on corner.

Roxy's Gang hit the airwaves from the broadcasting studio; ceiling grille concealed the organ pipes.

and keep it in our office. It was nice of you to do this. Won't you please accept the compliments of the theatre and be our guest some time, and come up and shake hands with us?" Pohl took Roxy up on his rather regally worded invitation, and was ushered into the Presence for a chat one afternoon, given a complimentary ticket good for some future performance, and turned over to "the General of the ushers" for a tour of the theatre from broadcasting studio to engine room. Roxy, for his part, put Pohl's ticket in a little black frame and hung it on the wall of his office where it stayed as long as he was in residence.

Herbert Pohl's guided tour was an eye-opener. While the auditorium was necessarily the most wondrous wonder of all the wonders of the Roxy Theatre (its 6,214 seats could shelter the entire population of Forest City, Pa., with room to spare for visiting relatives), there was a great deal more to the theatre building than met the eye of the ordinary visitor. Five floors of dressing rooms, served by two elevators, took care of the performers. There were the big chorus dressing rooms, with rows of mirrors outlined in light bulbs, and there were two whole floors of private dressing rooms, each with its own complete bath, for stars and principals. Each dressing room had a bed, a spacious closet, a writing desk, in addition to the makeup table; all that was missing was a Gideon Bible.

There was lodging for animals in a large room reached by a ramp just below stage level, and here were accommodated all kinds of beasts, from the anonymous camels and donkeys who took part in the Christmas pageant to Captain Proske's Tigers, Powers' Elephants, Pallenberg's Bears, Dr. Ostermaier's Wonder Horse, Alf Loyal's Dogs, Gautier's Toy Shop Ponies, and Sharkey the Seal . . . as well as a Noah's Ark full of other furry, finny, and feathered headliners through the years.

Above the dressing rooms was a floor given over to the broadcasting operation; the main studio was large enough to hold Erno Rapee and the 110-piece Roxy Symphony, the Roxy Chorus, and assorted soloists, plus Lew White at the two-manual console of a Kimball pipe organ whose pipes were installed in two chambers in a penthouse above the studio and played via a tone chute through grilles in the ceiling.

Often there were as many as two hundred people swarming over the studio when Roxy's Gang went on the air; in addition to the regular Gang and the musicians, there were guests—usually artists appearing on the theatre's stage for the week. The broadcasts were always lively and tumultuous, particularly on the occasions when Roxy would let Henry Heil step to the microphone as "guest announcer" for the station break. Henry was the former Rialto doorman from whom Roxy had borrowed four hundred dollars to help launch the ill-starred Rothapfel Unit Programme in 1918. When the Cathedral of the Motion Picture opened, Henry became its sexton, spending most of his time in an easy chair outside his beloved "Mr. Roxy's" office. He had a set of walrus mustaches and a comic German accent to match. The sight and sound of him gravely intoning: "Dis is der Plue Net-vork off der Nation-all Brroadcasting Cawmpanee" always had the studio in hysterics before the program could resume.

With Maria Gambarelli giggling in his ear, Douglas Stanbury had a hard time holding back his baritone laughter while singing, every week. But on one program he was given a song to sing that broke him up so badly (to say nothing of everyone else in the studio) that he finally had to turn his microphone to the wall in order to get through it. The song was called "All Alone With a Sandwich," and it was about Lindbergh. Admiral Plunkett's wife had written it specially for Roxy to use on the air, and nothing could persuade

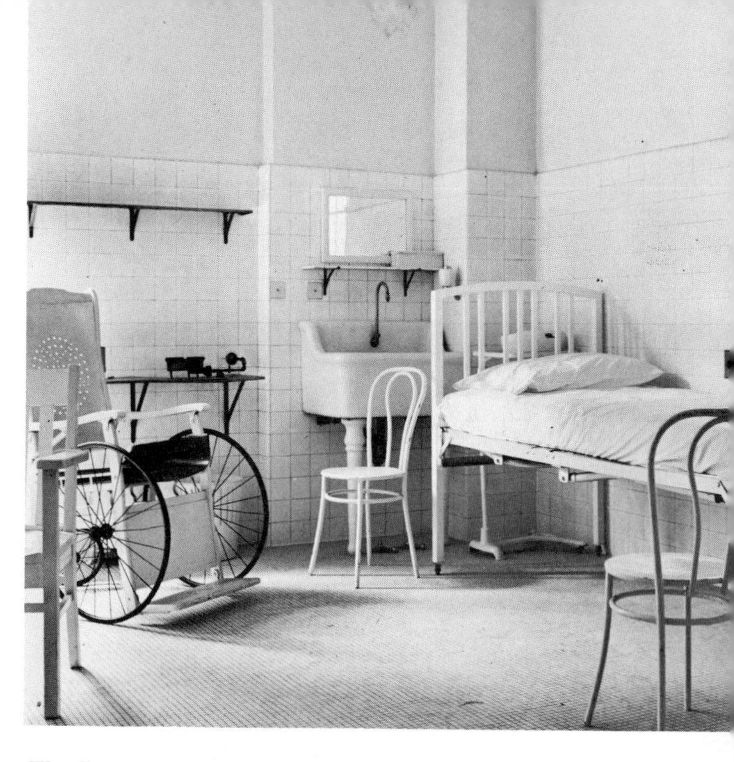

The Roxy Hospital was a 3-room sanctuary for the faint, the sprained, the tired — but not the *enceinte*.

Roxy to forget all about it. Not even the lines: "So, with one fell swoop/ He looped-the-loop/ And landed right in Paris." Major Rothafel, USMC, might risk offending some tycoon's wife, but an admiral's lady—never!

Off the main studio were more dressing rooms, a sizable rehearsal hall, and a glass control booth located on a mezzanine halfway up the wall of the high-ceilinged studio.

Also on this floor were entrances to the golden staircases that wound their way down into the auditorium on each side of the stage. And here were doors to tunnels that led to blood-chilling catwalks out over the stage itself, to the room housing the giant twenty-one-foot carillon chimes, and to the chamber enclosing the pipes of the Fanfare Organ, high above the stage on the right-hand side. One of these tunnels led to a spooky catwalk around the inside ledge of the dome over the auditorium, to the spotlight stations over the balcony, and here six operators perched behind their sizzling spotlights, following cues over the house intercom from the stage manager 180 feet below.

On another floor backstage was the ballet room, a barren and businesslike hall with mirrored walls and practice bars, a piano, a few chairs and nothing else. The costume department, looking like a loft in the garment district and filled with sewing machines and old ladies and sequins and chiffon, was on a lower floor, and below it was the dining room where a kitchen was kept busy from breakfast to midnight suppertime, feeding hungry Roxyettes who danced, kicked, and unicycled their way through five meals a day without showing a pound. In the floors below-stage were a laundry and dry-cleaning shop, a hairdressing parlor and barbershop. And there was a rehearsal theatre complete with stage and movie screen where an entire show was put together before being presented at dress rehearsal on the great stage in the theatre above.

Most of the area on the lower levels was painted a glistening gray, and the sound of the air-conditioning equipment, the electric generators, the organ blowers, and the various lift mechanisms heightened the impression that the whole place was actually somewhere five decks down in a battleship. Wherever the floor met the corner of two walls, there was a daub of white paint to reveal any stray dust or carelessly tossed cigarette butts. Red lines were painted on the floor of some of the tunnels to guide wanderers to safety through the maze that ran under the theatre. Here were the musicians' dressing rooms and shower rooms, their club room with its pinochle table and telephone booths, and here was the shoe-shine stand whose *padrone* had embossed the letters R-O-X-Y in its brass base. There were scene shops, upholsterers' shops, electricians' shops, and the master carpenter's shop. There were prop rooms containing dragons, drawbridges and droshkies, parasols, parachutes, and parallel bars.

There was one room filled with nothing but great wooden balls painted gay colors which the Roxyettes used from time to time to get about on instead of dancing. There was a unicycle garage. And there was a chamber whose walls were covered with racks containing font after font of milk-glass letters mounted in rectangular metal sheets; some sheets contained whole words like "The" or "with" or "and" or "Movietone!" This was the composing room of the marquee changers who every week would follow carefully plotted diagrams of the Roxy marquees to post the new attractions.

There were, by conservative estimate, sixty complete bathrooms in the Roxy, 212 flush toilets, forty-five shower stalls. And this doesn't include Roxy's personal health club in his penthouse suite high above the Rotunda. He had a massage table, a sun lamp, and a steam room; his shower bath had special needle jets coming from all angles, and a flexible shampoo shower head. His dressing room had enormous cedar closets where he kept a part of his wardrobe of eighty suits. It also contained a rack of Indian clubs. He had a dining room that would seat fourteen, with a kitchen to match (and here hot dogs were prepared, not Forest City or Rivoli style, but in silver chafing dishes). There were two bedrooms (often Roxy didn't get home to Rosa and the children for days at a time). And there was, of course, his office.

An outer office guarded the double doors, and an elaborate system of buzzers and annunciators heralded visitors, but often as not Roxy would bound out to greet his guests before the secretary could get through the required formalities. Dominating the office was a giant Arthurian fireplace carved of limestone; on the south wall were three arched windows with panes of amber glass. The walls were paneled in walnut and the floor was carpeted in what looked like unborn lamb. A giant desk six feet across, almost six feet wide, filled one corner. On its glass top were two telephones, a pipe rack, a pen stand, a humidor, a picture of Rosa and the children framed in silver, and some books between marble bookends: a dictionary and the complete six-volume set of *Human Nature and Practical Salesmanship.* No *Stoddard's Lectures.*

Beside a couch upholstered in wine-colored plush was a grand piano. One of Roxy's jokes was to wander over to the piano during an interview, start playing an intricate bit of Chopin or Liszt with a great deal of shoulder dipping and body English, then get up and light a cigar while the music continued to play. It was his Aeolian Duo-Art—an electric player piano—and one of his favorite toys. A tunnel and a winding staircase led from the office to Roxy's private box, overlooking the balcony in the auditorium. It resembled nothing so much as Nero's box at the Coliseum, with a canopy of crimson velvet fringed in gold, flanked on either side by twisted golden columns. Here he would entertain visiting caliphs and show-business potentates, unobserved by the thousands of "his people" below.

On the floor below Roxy's office was a projection theatre seating one hundred people; there was also a row of large and elegantly appointed offices for his staff — managing director, house manager, publicity director, booking agent, business manager, purchasing agent, and *their* staffs. And on the floor beneath this was the accounting department and the publicity shop with its files of photographs, mimeographed press releases, piles of scrapbooks. The next floor was the domain of the ushers, an establishment that was part Marine boot camp and part YMCA.

The below-decks shoeshine stand had "Roxy" hammered into its brass floor by its devoted attendant.

There were club rooms, with a billiard table and magazine racks; a small gymnasium; a uniform shop; a cafeteria for off-duty snacks; and dressing rooms and showers. There were also bunk rooms where weary ushers could catch a nap between shifts. The orderly room where the Chief Morale Officer (a retired Marine colonel) sat was a model of Spartan spit-and-polish, and here duty rosters, "gig" lists, citations and notices of promotion were posted in a most military manner. Being a Roxy usher was a serious business and there was no room in the corps for weaklings.

Below the ushers' garrison was the Roxy hospital. Three gleaming rooms filled with everything from autoclave to tongue depressors were dedicated to maintaining health not only among the Roxy staff and the Roxy patrons, but the whole neighborhood around Seventh Avenue and Fiftieth Street. According to a press release, in the first year after the theatre opened nurses Grace Marigla and Anne Beckerle treated 12,900 individual cases ranging from headaches to skinned knees (belonging to the Rothafel children) to emergencies brought in from outside the theatre. While the Roxy hospital took credit for saving eight lives that first year, there was not (nor was there ever in the theatre's history, contrary to popular legend) a single *accouchement* accomplished in the Cathedral of the Motion Picture.

•

Getting this bastion of glamor, efficiency, technology, and dreams ready for opening night required the twenty-four-hour services of an army of plasterers, upholsterers, glaziers, drapers, electricians, carpenters, and carpetlayers. The last week was something that Harold Rambusch and his decorators can never forget. Whirling through

The Roxy balcony seated 2,320; recess for projection booth was at front, Roxy's private box was at top, spotlight gallery was in dome above.

Orchestra lift in "overture" position (left) showing seats for 110-man Roxy Symphony, 3 organ consoles in front. View from stage (far left) shows intricate golden staircase and pulpits, used by soloists, chorus and ballet during services in Roxy's Cathedral.

all the confusion, like a tornado in a circus tent, was Roxy himself — screaming at contractors, soothing musicians, countermanding his own orders while they were still echoing from the crags of the balcony. One day remains particularly memorable. Hammers were crashing through imported mirrors, paint was dribbling down tapestries, fuses were blowing like popcorn. Down on the stage a baritone was singing, "... just a Rus-sian lullaby," while the corps de ballet was prancing a disorganized *pas-de-trente-et-deux* to the beat of an exhausted rehearsal pianist. High above this Donnybrook, on a scaffold cantilevered out over the balcony, Rambusch was trying to soothe a crew of edgy plasterers as they gingerly surfaced the last section of the dome.

He remembers hearing Roxy shout from below that he was coming up with a visitor. Up the ladder and out the swaying catwalk teetered Gloria Swanson, followed by the impresario himself. Roxy explained that since the theatre's opening film was going to be Gloria's new picture, *The Love of Sunya,* he was giving her a personally conducted tour of the premises. Then, before Rambusch could say anything, Gloria danced out past the dumfounded plasterers, picked up a mason's trowel, and scratched:

"Dear Roxy—I love you—Gloria," in the wet plaster of the dome.

Roxy was so touched by Gloria's sentimental prank that he gave instructions that it should be carefully gold-leafed and left there forever.

"A Cavern Of Many-Colored Jewels"

There was a third titan in the field of American movie palace design: sharing the honors with John Eberson and Thomas W. Lamb was the Chicago firm of C. W. & George Rapp. The Rapp brothers got off to an early start when Balaban & Katz commissioned them to design their first theatre, the Central Park (see page 202) in 1916. From that day forward, until the last movie palace was built, their drawing boards were never without at least one major project for either Balaban & Katz or for Publix Theatres (Same Katz's subsequent triumph and the biggest chain of them all).

While Eberson went in for breathtaking meteorological effects, and Lamb ran the gamut from staid Adam elegance to Arabian Nights phantasmagoria, Rapp & Rapp put one idea above all others: eye-bugging opulence. They knew (just as the others did) what the public wanted in its movie palaces, but their theatres offered escape not into a world of starlit gardens or double damask dignity or temples of Vishnu; their stock in trade was a grandeur that spelled m*o*n*e*y to the dazzled two-bit ticket holder. In an article in 1925, George Rapp summed up his thesis in terms that put him right in the same architect-of-words class with Eberson and Lamb:

> Watch the eyes of a child as it enters the portals of our great theatres and treads the pathway into fairyland. Watch the bright light in the eyes of the tired shopgirl who hurries noiselessly over carpets and sighs with satisfaction as she walks amid furnishings that once delighted the hearts of queens. See the toil-worn father whose dreams have never come true, and look inside his heart as he finds strength and rest within the theatre. There you have the answer to why motion picture theatres are so palatial.
>
> Here is a shrine to democracy where there are no privileged patrons. The wealthy rub elbows with the poor — and are better for this contact. Do not wonder, then, at the touches of Italian Renaissance, executed in glazed polychrome terra-cotta, or at the lobbies and foyers adorned with replicas of precious masterpieces of another world, or at the imported marble wainscoting or the richly ornamented ceilings with motifs copied from master touches of Germany, France and Italy, or at the carved niches, the cloistered arcades, the depthless mirrors, and the great sweeping staircases. These are not impractical attempts at showing off. These are part of a celestial city—a cavern of many-colored jewels, where irridescent lights and luxurious fittings heighten the expectation of pleasure. It is richness unabashed, but richness with a reason.

In the dream world of the movie palace, everything was twice as rich, three times more fanciful than life. Colors were brighter than life, too; under amber or violet auditorium lights a delicate pastel mural would look as washed-out as a daguerreotype. An architect usually made a color rendering of a theatre interior to show to his client before any blueprints were drawn. On the following pages are a few examples of renderings in "glorious color" that show how much the architecture of pleasure depended upon paint, gold leaf and imagination.

Rapp & Rapp theatres lifted the spirits of movie goers all over the country, but the greatest concentration of them was, naturally, in Chicago. The Tivoli, that towering Versailles on Cottage Grove Avenue; the Chicago Theatre, pride of the Balaban & Katz fleet; the Uptown, with its 46,000 square feet of Spanish-type grandeur; the Norshore, a posh 3,000-seater on Howard Street; the Palace, a Louis XIV salon dedicated not to B & K but to B. F. Keith; the Picadilly, a lush neighborhood house for H. Schoenstadt, are a few of the more outstanding. And there was the Oriental — Rapp & Rapp's one flagrant excursion into the gimmicked world of Scheherezade, decreed by Balaban & Katz in the Masonic Temple Building (also Rapp & Rapp designed) on Randolph Street.

The Oriental was described in a Rapp brothers' press release as "an educational treat in itself as a work of art . . . to study and examine the array of sculptured detail throughout the theatre is like a trip to the Orient. The auditorium is beyond description with its intricacies of Eastern magnificance, grotesque dancers and Indian animal figures, resplendent with lights behind colored glass around ornate shrine-like niches."

Rapp & Rapp can be forgiven the Oriental when one considers the tasteful splendor of the Paramounts in New York and Brooklyn; the Michigan Theatre in Detroit with its stupendous canopy and vertical sign looming over Bagley Avenue; Rubens' Rialto Square Theatre in Joliet, Illinois, a giant theatre in a small town that attracted patrons from all the surrounding counties; Shea's Buffalo Theatre, with its golden stage gates similar to the Uptown in Chicago; and that unique gem, the Al Ringling Memorial Theatre in Baraboo, Wisconsin, a gold and red plush replica of a tiny European opera house.

The Oriental Theatre in Chicago was designed by Rapp & Rapp as a showcase for Paul Ash and His Merry, Mad Musical Gang, but the frenetic Ash must have had a tough time competing with the Oriental's hasheesh-dream decor. Even the Mighty Wurlitzer arose emblazoned with crimson fire birds. The Oriental sits vacant on Randolph Street.

Rapp & Rapp branched out into the field of non-theatrical architecture with such notable structures as the Corn Palace in Mitchell, South Dakota, which was ornamented from foundation to roof with stalks, sheaves and ears of corn. There was the Bismarck Hotel and the palatial Windemere Hotel, both in Chicago, and the Leland in Detroit. There were Sigma Chi fraternity houses in Champaign, Illinois and Oxford, Ohio. There were banks, auto showrooms, office buildings and churches. And, final proof of their status as "architects by appointment to Balaban & Katz," the Balaban Mausoleum in Chicago.

The Ambassador Theatre in Saint Louis was designed by George & C. W. Rapp of Chicago in 1928 and heralded by coming of *arte moderne* to the movie palace. They specified silver leaf everywhere, even on the traditionally golden Mighty Wurlitzer. The jewels on the organ screen twinkled electrically. The Ambassador, long derelict, is threatened with demolition.

The Uptown in Chicago, by Rapp & Rapp, had golden gates that opened on the silver screen or the Balaban & Katz stage shows. Note dials on conductor's desk in orchestra pit; he could control the speed of the projectors in the booth if the silent movie was not keeping up with orchestra. The Uptown, empty and heavily water damaged, still stands.

The Los Angeles Theatre, designed by S. Charles Lee, had a crystal fountain at the head of its grand staircase (right), illuminated glass strips in the aisle floors, a prism device in the downstairs lounge for viewing the show in the auditorium above, electric cigarette lighters in the walnut paneling of the retiring rooms (the Gents had a pink Carrara marble shoeshine stand). Its faded grandeur is still seen on South Broadway in Los Angeles.

Paramount
Paramount
POLA NEGRI IN THE WOMAN ON TRIAL
PARTINGTON'S "FLYIN' HIGH" WITH
BEN BLACK AND PARAMOUNT BAND
RIDE'S TICKETS
MIDNITE PICTURE
HATTER
DRESSES Beverly MILLINERY

The marquee and vertical sign of the Atlanta Paramount, nee Howard (right; also see page 24) was typical of the hundreds of Paramounts that dotted America (the Atlanta theatre was razed in 1960). But the *paramount* Paramount was built on Times Square in New York. The most graceful, though not the most flamboyant, of all of Rapp & Rapp's theatres, its lobby was inspired by l'Opera in Paris (far left), and there was statuary to spare (left) in addition to a gallery of romantic oils, a Hall of Nations with stones from all over the world, and a *salle de musique* on the mezzanine where a Fragonard-embellished grand piano was played to entertain waiting patrons in the lobby below. The arch of the great window above the handsome street canopy (see preceding page) was repeated on the proscenium in the ivory, green and rose auditorium. The Paramount, long a bastion of elegance on tawdry Times Square, was gutted for office use.

The Brooklyn Paramount, built three years after New York's, was by Rapp & Rapp in a fancier mood. The sunburst effect of the proscenium (right) was enhanced by the semi-atmospheric latticed ceiling over the balcony (above). The vine-drenched arches along the walls (see following page) were lit by Wilfred "color organ" that painted moving shapes in colored light, an inspiration of Frank Cambria who decorated the theatre. Curving staircases in front or the organ grilles led chorus girls – among them, Ginger Rogers – stageward.

The Fox in Atlanta was designed in 1929 by Marye, Alger & Vinour to be admired from outside as well as inside. Its minarets and domes towered over the old Tompkins mansion on Peachtree Street (upper right) and its bronze marquee faced not Mecca but the Georgian Terrace Hotel across the way. Cream and buff brick, set in ribbon-like courses, ran all around the theatre (right); the fire escapes got full architectural treatment, ascending in terraces up to the bronze dome surmounting the mosquelike side entrance. The organ grilles in the auditorium (left) were lit by changing colored lights when the giant Möller De Luxe played.

FOX
WILL
ROGERS
TALKIE

EXIT
EXIT

The atmospheric auditorium of the Atlanta Fox (left) seated 5,000 in an open courtyard surrounded by castellated walls of cast stone. The box office (above) repeats the Alhambra-inspired motif of the organ grilles. The retiring rooms were Egyptian, Moorish and Turkish delights, furnished by the peripatetic Mrs. Fox after one of her rummaging tours through the bazaars of the Levant. The Fox remains as Atlanta's principal showplace, is host to the Metropolitan Opera for a week's season every spring.

Grauman's Egyptian Theatre, Meyer & Holler, architects, brought Karnak to Hollywood Boulevard. The forecourt (above) was flanked by a bazaar of shops; on the parapet a Bedouin in striped robes chanted the title of the feature. The stage (left) made King Tut's tomb look like the old family burial vault, but the rest of the auditorium (right) as seen in this view toward Sid Grauman's private box was strikingly simple for 1922. The Egyptian is still successful as the home of long-run epics.

Grauman's Metropolitan Theatre (above) might well have been named Grauman's Persian. It was designed by William Lee Woollett, who made novel use of cast concrete in decorative forms. Theatre opened in 1923, was renamed the Paramount in 1927, and in 1960 was demolished. Mae Murray (right), in a happier time, feels the mystery of the Metropolitan's ornate lobby.

Grauman's Chinese (above) opened in 1927 shortly after the Roxy and immediately became *the* Hollywood movie palace. It was designed by Meyer & Holler, architects of the Egyptian. The auditorium (see next page) had walls of painted brick, columns of cast concrete, and a dazzling ceiling lantern that concealed the organ pipes. As a signal that one of Sid Grauman's famous prologues was to begin, the gong hanging between the columns was struck. A larger gong, high over the pagoda box office (upper right), summoned audiences from the forecourt where they had been studying the footprints (center right) and autographs of Hollywood stars. Wax figures garbed as mandarins and sing-song girls guarded the lobby (lower right); Fu Manchu & Co. are still on duty.

Greetings to Sid
1927
Mary Pickford
HAND AND FOOT PRINTS OF

The Fifth Avenue Theatre in Seattle very nearly out-Chinesed Grauman's Chinese. The monstrous golden dragon in the ceiling "grasping for the white globe suspended below, symbolic of the Pearl of Perfection," is still grasping.

"TAKE THE GRAND STAIRCASE TO THE LEFT"

"We, the attachés of the Roxy Theatre, earnestly request our patrons to kindly refrain from offering gratuities for any services rendered. We have pledged Mr. S. L. Rothafel, "Roxy," that we will under no circumstances accept payment from his patrons for courtesies we enjoy extending to them. We regard the Roxy Theatre as a university and place ourselves in the position of students seeking better understanding and appreciation of theatre arts. Patrons of the theatre are our guests and we place ourselves in the position of hosts. The offering of a gratuity will be mutually embarrassing because it will politely be refused. Being associated with Mr. Rothafel is a distinct privilege and pleasure that we feel is sufficient remuneration."

THE ATTACHéS
COL. HOWARD H. KIPP, USMC (RET.), *Morale Officer*
CHARLES F. DOWE, *Chief Usher*
(From the first *Roxy Weekly Review,* March 12, 1927)

Major Rothafel's Dragoons, phalanxed in the Rotunda.

There was something about a movie-palace usher that was fine, fine, fine. Somehow taller, more stately and omniscient than ordinary mortals, he inspired confidence in old ladies, heart throbs in flappers, and fear in such potential trouble-makers as Bronx cheerers in the balcony or mashers in the mezzanine.

The staff of a well-operated theatre included an average of two ushers per aisle per shift (usually two daily) who reported to a chief usher or floor captain. In addition, there were page boys to deliver messages, direct visitors, and help in the check room; elevator operators (if necessary); a doorman—a personage of great importance, tact, and smiling countenance — who checked for drunks, Pekinese, and gate crashers while guarding the portals; streetmen who supervised crowds in the outside lobby and sidewalk and who, in some instances, would relay a patron's ticket request to the lady in the box office with mysterious finger signals; footmen who opened car doors (with a genial half-salute), held an umbrella over dismounting patrons in rainy weather, and shooed away cruising taxis, but never, never, under any circumstances, touched a patron unless asked to. Footmen always got to wear exotic uniforms, and if occasionally they showed tendencies toward gigantism, so much the better.

There were also a number of ladies on the staff, even in a theatre that didn't use usherettes (and not many did). There was the cashier in the box office, who had to be blonde, love roses, but "not be expected to make any unnecessary sacrifice in case of holdup or robbery"; the nurses presided over the hospitals or first-aid rooms and were trained in midwifery only to the point of knowing

The ushers at the Avalon in Chicago, in kepis and white gloves, seem ready to join the French Foreign Legion — or maybe it's the cast of *Beau Geste.*

how to gently but rapidly hustle any imminent mother-to-be out of the theatre and into an ambulance at the first hint of labor; matrons held forth in the retiring rooms and were combination beauticians, seamstresses and mother confessors to girls who sought the refuge of the Ladies after an emotional upset in a back-row seat. Other motherly women patrolled the kiddie sections of the orchestra during matinees or doled out Tinker Toys in the nursery.

But it was the ushers who were the pride of the theatre staff. Understandably, they developed a remarkable loyalty to their uniform and to its traditions. Only the higher types of boys were hired, and many theatres showed a genuine responsibility towards their welfare. (At Loew's Kings Theatre, on Flatbush Avenue in a well-to-do section of Brooklyn, there was a fully equipped basketball court installed beneath the lobby, and here teams from Loew's Pitkin or Loew's Kameo or even the Brooklyn Paramount might challenge the home quintet.) Being an usher was considered a highly attractive job for any boy, and the prospect of seeing all those movies free made it all the more sought-after. Recruits were trained in the arts of Emily Post and General Pershing, and they were, in many instances, like the Roxy ushers who placed themselves "in the position of students seeking better understanding and appreciation of theatre arts."

Being a Roxy usher was like being a West Point Cadet; there was just nothing higher for a boy to aspire to than to be a member of that elite guard—hand-picked for manly bearing, devotion to duty, and freedom from acne.

A contemporary account in *The New Yorker* called them "young men of far greater beauty and politeness than any other clique in the city; it is impossible to describe the Roxy ushers—you have to go there and be pampered by them to know what sweetness really is." A 1927 joke in *Judge* went:

"I think the portieres at the Ritz are gorgeous!"

"But, my dear, have you seen the ushers at the Roxy?"

And later Cole Porter immortalized the boys in a line from "You're the Top" —

. . . you're the steppes of Russia.
You're the pants on a Roxy usher.

The life of a Roxy usher was surprisingly complex. The first battalion reported for duty at ten in the morning in the ushers' quarters on the fourth floor, where civvies were changed for uniforms. Then a rigid locker inspection was carried out by Colonel Kipp, the Morale Officer. There were forfeits in order for the boy who was late, whose locker was untidy, or whose uniform needed repairs.

At 11:10 a bugle sounded Adjutant's Call and Sgt. Gene Le Gendre, an ex-Marine DI, marched them in a column of twos through the door of the ushers' quarters, down the sharply curving stairs of the Tower Foyer at top-balcony level, across the balcony promenade, down the Grand Staircase and into the marble-columned drill field of the Rotunda. Here orders of the day were read by Chief Usher Dowe, citations and promotions were announced, and flashlight batteries were checked. This formation was followed by the daily fire drill and first-aid instruction at 11:15. At 11:30 Drill Master Le Gendre put them through some intricate close-order drill. At 11:35 the bugle sounded Tattoo, the house lights came on, and the ushers were marched off to their respective posts.

The usherettes at the Palace in South Bend, Indiana, celebrate a rollicking fifth anniversary in the lobby, well-chaperoned by their no-nonsense Matron.

Each usher was responsible for inspecting his own territory and should anything be wrong, such as a seat that wouldn't work, a carpet that was snagged, or an aisle light out of order, he made a report to the Chief Usher who notified the duly constituted authority on the Roxy maintenance staff.

At 11:45 the lobby doors were opened. The bugle sounded First Call as the ushers snapped to attention and the Cathedral of the Motion Picture —manned by Roxy's answer to the Swiss Guard —was ready for business.

Each evening at six, as the Rotunda pipe organ tootled martial airs, they were marched, 125 strong, into the Rotunda again by the Morale Officer and his aides-de-camp.

"RIALTO" HAS GYM FOR STAFF

Manager Robt. Weitman of the Rialto Theatre, New York, used ingenuity for money and now the staff is always full of Pep. Here are some of the ushers charging their pep batteries.

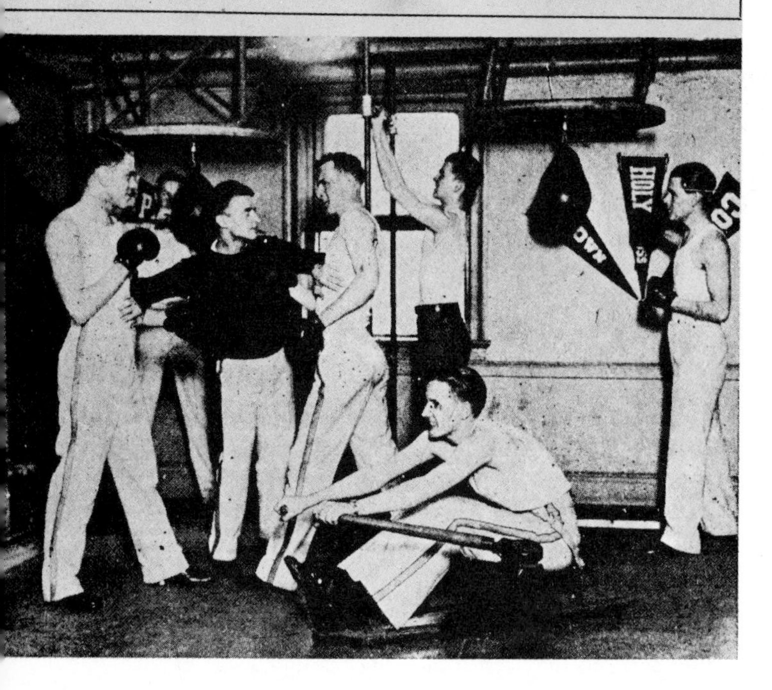

After executing a number of spectacular maneuvers with an insouciant precision rivaled only by their onstage auxiliary, the Roxyettes, the daytime ushers, in their smart dress blues, surrendered their flashlights and emergency pouches to the evening ushers in white tie and gold-braided mess jackets. The emergency pouches contained smelling salts, pad and pencil for messages and accident reports, a spare pair of clean white gloves, and the usher's own individual tin of Sen-Sen which had been issued by the Roxy Quartermaster in much the same way as prophylactic kits.

The buttons on their uniforms all bore the same "R"-monogrammed microphone motif as the oval rug they marched upon; their shoes (which were equipped with special steel arches) glistened like limousine fenders, and their faces shone with goodness. When the Changing of the Ushers was over, patrons, who had thronged the Grand Staircase to witness the ceremony, could go to their seats secure in the knowledge that the Roxy's ramparts were being watched by the brave and the true.

Though Congressional appointment was not actually necessary, getting onto the Roxy uniformed staff was a complicated business. Each week dozens of boys tried to enlist in the corps; only a few of the more prepossessing were chosen for training. After being screened for charm, poise, no fallen arches and a haircut, the recruit

Troops deployed at left are members of the Publix militia on duty as palace guard at the New York Paramount. The Roxy ushers (below) are demonstrating their "manual of arms" – a private sign language that could locate and enumerate the best remaining seats, put the finger on a purse-snatcher or warn comrades of approaching Roxy brass.

was sent to military school on the fourth floor for a week. Upon passing requirements worthy of Culver, the recruit was next trained in etiquette—he must be able to shade his voice and keep it always cultured, manly, refined, pleasant and optimistic—the last without even the scent of a tip. In signaling a patron (or a fellow usher) he must be quiet and dignified, *never* snapping his fingers. "Hey, you," would put him on cafeteria police for a week. Pointing was proscribed; directions to the ladies' room were given in appropriately confidential tones—"Under the arch at the left of the Rotunda, madam, then down the staircase and to your right."

Patrons at the Roxy were sometimes a bit overawed by these splendid young men, and no wonder. The boy whose mother raised him to be a Roxy usher had to live by a code of conduct that brooked no monkey business. There were certain misdemeanors that called for discharge other than honorable: accepting tips meant Taps, but in declining, the usher must not embarrass his would-be benefactor. And he must never flirt on duty; the Roxy manual revealed that there were many ways of flirting—(1) by use of eyes, (2) by use of speech, (3) by action—and all these methods were forbidden. Ushers who got something in an eye went on sick call rather than risk being caught winking in the loges.

The ushers, of course, had an elaborate system of communication that was not unlike the sign language used by deaf mutes. A message from the Chief Usher ("An old lady in a wheel chair heading for Aisle 11—check her brakes." Or, "Jiggers—here comes Roxy through the Rotunda with some big shots!") would be transmitted all over the theatre with the speed of Zuñi smoke signals.

Being a Roxy usher was a way of life. Every evening at six they were marched down to the Rotunda for the "changing of the ushers" ceremony (right), and there were rousing checker games in the day room on fatigue time.

Many movie-palace managers, in organizing their own usher training along the lines of Major Rothafel's Academy, made the mistake of encouraging their boys to salute patrons and click heels at every turn. Roxy would not allow his patrons to be thus intimidated, yet recognizing that an *esprit-de-corps* such as his boys had must have some outlet, he permitted the ushers to salute each other, as well as any Roxy executive they could catch in the theatre. As for heel-clicking, that was for flamenco dancers.

Besides the "privilege and pleasure" of being associated with Mr. Rothafel, there were other distinct advantages to being a Roxy usher. Education was encouraged, and college boys were given a high priority on the recruiting lists, particularly if they participated in ROTC. Roxy provided (in addition to the clubroom and library) a study hall on the fourth floor, and many a term paper was written there between shifts in the upper balcony. There was even the Roxy Fraternity Club to which only the bravest and truest might be tapped for membership. Each summer the boys were packed off to camp in the Catskills, a platoon at a time, and those who stayed behind could be seen any morning on the bridle path in Central Park being tutored in the art of horsemanship by Sgt. Le Gendre. Just how Roxy planned to utilize his cavalry in handling the crowds in the Rotunda was never explained.

Roxy's paternalism extended in many ways, and it was a genuine fatherly sentiment. The man whose own father had "chased him out of the house" at the age of fourteen knew the value of guidance and affection in shaping futures; he never had it. That his ushers all had futures ahead of them was one of his most cherished and sentiment-washed beliefs.

"It is not foolish to assume," he told a reporter not long after his theatre opened, "that many captains of industry of the future will hark back to their early training at the Roxy Theatre and say, as they sit behind their desks, 'God bless you, Roxy.' "

During Anniversary Week they appeared onstage, supported by their Auxiliary Corps, the Roxyettes.

Sometimes the Glee Club gathered to sing the Roxology.

And there were chances for heroism in line of duty.

Some of New York's eminent conductors in 1921.

". . . THE PEAL OF THE GRAND ORGAN, THE FLOURISH OF GOLDEN TRUMPETS"

"To demonstrate the remarkable talent of my musical staff, I take pleasure in presenting them in solo work."

Advertisement for The Family Theatre,
Forest City, Pa.—1908

Samuel L. Rothafel — and the American movie palace — had traveled a long road between Miss Mabel Rennie at the upright in the Family Theatre to Erno Rapee and the 110-man symphony in the Cathedral of the Motion Picture. But Roxy never faltered along the way in his musical taste (God-given, it must have been; he never studied music and he certainly didn't inherit his love for it from Gustave Rothapfel, the shoemaker of Stillwater). In the mid-Twenties, when other showmen were turning from serious music to jazz, Roxy was steadfast in his belief that the best thing to offer moviegoers in the biggest theatre in the world was something they couldn't get in Forest City or Los Angeles or the Bronx.

"When we undertook to have the Capitol Grand Orchestra play Richard Strauss' symphonic poem *Ein Heldenleben*, several years ago," he observed, "we embarked on a daring adventure in the field of popular entertainment. The overwhelming success of its reception by our audiences was the most gratifying and encouraging element in our performance of this composition, and it justified our belief that our audiences are the finest in the world."

The same week in 1922 that Roxy presented *Ein Heldenleben*, an old Rothafel alumnus, Dr. Hugo Reisenfeld (who had taken over the direction—both musical and managerial—of the Rialto and the Rivoli when Roxy left) demonstrated that even the most serious tastes falter in the face of popular demand. Ads for the Rialto proclaimed "Reisenfeld's Classical Jazz—the Famous Rialto Orchestra proves it can 'get up and go.' "

Erno Rapee joined Rothafel at the Rivoli, later rejoined him at the Capitol, and when the Roxy Theatre opened, became its musical adviser as well as Director of the Roxy Symphony Orchestra. He was assisted at the Roxy by conductors Charles Previn, Frederik Stahlberg, and H. Maurice Jacquet, all musicians of both reputation and skill. Working closely with them were Leo Staats, *Maitre de Ballet*, and Leon Leonidoff, Associate Ballet Master (who soon deposed the *maitre* when Staats's French proved too difficult for the young ladies of the corps).

The Roxy Symphony took itself seriously, and soon proved that it was a musical organization of the first order. The Roxy Sunday Concerts (presented at 11:30 in the morning as a way to get non-churchgoing patrons into the Cathedral during church hours and still comply with New York City's strict Sunday ordinance against movies during worship) began to attract serious music-lovers as well as Sabbath-breakers. Wagner, Richard Strauss, Bloch, Mozart, and Saint-Saëns were program favorites with the Roxy Symphony, and it was quite an experience to settle down for a prenoon concert of Brahms and Borodin with the delicious knowledge that when it was over there awaited the complete Roxy de luxe performance, Lew White, Gamby, The Russian Choir, The Roxyettes and all. And a movie, too.

Every town had its local favorite among moviepalace orchestra leaders. At the Allen Theatre in Cleveland, it was Philip Spitalny, long before his all-girl period; in Los Angeles, Carli D. Elinor, at the California Theatre, was considered without peer; Atlanta's Enrico Leide, who had come there to open the Howard Theatre in 1921 from an engagement at the Colony Theatre in New York, rose to glory with the Atlanta Fox Symphony, led a stage jazz band at the Atlanta Capitol during the Depression, and went on to make a new name for himself as a conductor of opera and concert works in later years.

These were only a few of the *maestri* who made good as local fixtures. In New York, it seemed there were more fine musicians in the pits of movie palaces than there were in all the opera houses and concert halls in the rest of the nation. In 1926 the Capitol Theatre, carrying on the musical tradition so emphatically established by Roxy during his tenure there, could boast an eighty-piece orchestra conducted by the redoubtable David Mendoza, assisted by Eugene Ormandy as

Dr. Hugo Reisenfeld, as guest conductor, takes a bow with the Roxy Symphony. The orchestra, only partly shown here, are in their white summer jackets. At the organ are Alexander Richardson and Dr. C. A. J. Parmentier.

In 1927 the Capitol Theatre ran a series of personality cartoons on the backs of the weekly programs. David Mendoza was conductor of the Capitol Grand Orchestra and Eugene Ormandy was associate conductor; their musical personalities were as different as their haircuts.

Associate Conductor. Ormandy in later years credited his days in the Capitol as having great value in readying him for his ultimate career. "Works were played by the week," he recalls, "and this meant that each one got performed twenty-eight times, Tchaikowsky's *Fourth,* Beethoven's *Fifth,* or whatever. By the end of the last show on Saturday night, you *knew* that music."

At the Mark Strand, Carl Edouarde led the Strand Symphony, and his taste was every bit as good as Mendoza's or Rapee's or Reisenfeld's, even if his orchestra was not as large or as highly budgeted. At the Paramount Theatre the General Music Director was Nathaniel Finston, assisted by Irving Talbot, while the concertmaster of the Paramount Orchestra was Frederik Fradkin, the noted violinist. Three orchestrators, three music librarians, and a vocal coach rounded out the Paramount's impressive music faculty.

During Roxy's temporary absence from the theatrical scene while the Roxy Theatre was being built, Major Bowes made musical hay at the Capitol in anticipation of the competition that lay ahead when the Roxy was finished. A measure of his success is found in this plaudit by Paul Morris, music critic of the New York *Evening World:*

"Last night I went to the Capitol Theatre, where one of the most elaborate of musical entertainments garnishes the feature film. It is really extraordinary the amount of pains that are taken here to put over the music. The Capitol Orchestra is admirable. It is as large as any of the regular symphonic organizations and better than many of them. But movie fans come to be shown, not merely to be told. So there are dancers to interpret the music, and good ones, too. This week the overture was *Die Fledermaus,* and then they introduced Carlo Ferretti who gave us *Lolita's Serenade* by Buzzi-Peccia. Julia Glass, an excellent pianiste, performed Rubenstein's *Concerto in D Minor.* The Capitol Ballet Corps presented a 'Bal Masque' featuring Mlle. Desha, Miss Doris Niles and Miss Alice Wynne, and Mr. Chester Hale, the Capitol's ballet master, can be justly proud of the product of the ballet school he conducts backstage. After the feature film, a frivolous item called *Soul Mates*—one of those Elinor Glyn things—had been given a delightful scoring by the Capitol Orchestra, Dr. Melchiorre Mauro-Cottone, at the console of the Capitol's truly grand organ, rendered 'Impressions of Cesar Franck' in a most admirable fashion."

Scoring the pictures was a real art, and two of its greatest exponents were Erno Rapee and Dr. Hugo Reisenfeld. Between the two of them they turned out the scores for dozens of films, not only in the theatres where they were engaged, but for most of the other major Broadway houses as well. Frequently a score for a film's Broadway presentation by one of these masters would be used all over the country wherever the picture was shown. Reisenfeld's score for De Mille's *King of Kings* at the Gaiety Theatre on Broadway was also used when the film appeared on the inaugural program

George Olsen and his orchestra give out with some snappy razz-mah-tazz at the Los Angeles Paramount.

of Sid Grauman's Chinese Theatre in Hollywood. In setting this giant Biblical epic to music, Doctor Reisenfeld prescribed Tchaikowsky's *Pathetique* and the Holy Grail motif from *Parsifal*, with liberal borrowings from the "Hallelujah" chorus. And not to lose the common touch, he laced the whole thing with such popular hymns as "Lead, Kindly Light" and "There is a Green Hill Far Away."

Erno Rapee, a composer of popular music in his own right, introduced his "Diane" into the score for Fox's *Seventh Heaven* which was played by a live orchestra during its first-run engagement on Broadway and later recorded (by the Roxy orchestra) on the Movietone sound track that accompanied the film in less pretentious theatres. But most of Rapee's scoring was done patchwork, in the accepted movie-house fashion, with snippets of this and that to fit the mood of each scene. The Roxy Theatre's music library, from which Rapee drew his inspiration, was one of the most complete ever assembled. Its nucleus was Victor Herbert's own collection, purchased by Roxy from Herbert's widow in 1927. It contained many Herbert manuscripts as well as complete scores of symphonies, operas, and musical comedies. Roxy's librarians added to it constantly through the years, keeping it always up to date with the latest popular music as well as serious works as they entered the Roxy repertoire.

"Every big feature picture is made up of themes," said Rapee. "It is the job of the musical director to embody each theme in his score and combine them so perfectly that the melody slides gracefully from one to another. In my work I first determine the geographic and national atmosphere of the picture, and then I figure out the more important characters. There must be a musical theme for each character, and one for the entire scenario.

"Villains are particularly difficult," continued Rapee, "musically speaking. Ordinarily, any agitato will express a villain, but you must make a distinction between sneaky, boisterous, crafty, powerful and evil-minded villains. A crafty villain, who does not prove himself a two-fisted he-man at the same time, can be described by a dissonant chord being held tremolo, and very soft. If the villain happens to be of the brute sort, a fast-moving number would be more apt."

Erno Rapee's thesaurus of villainy, romance, passion and piety, *Moods and Motives for Motion Pictures* was a fat, thumb-indexed volume that became holy writ to pianists, organists, and leaders of small orchestras in theatres that were not large enough to maintain their own musical arranger and librarian. With it, the musician could score an entire picture simply by using Rapee's cross-index of suitable themes drawn from every composer from Moussorgsky to MacDowell.

Rapee never lost sight of the real mission of the movie-theatre orchestra: to set a mood, almost subconsciously, in the minds of the film-watchers. There was plenty of time for orchestral flash and spectacle during the stage show; music for silent pictures should enhance, not distract. "If you come out of the theatre almost unaware of the musical accompaniment to the picture you have just witnessed, the work of the musical director has been successful. Without music the present-day audience would feel utterly lost. With it they should obtain an added satisfaction from the show, and still remain unconscious of the very thing which has produced that satisfaction."

Louis F. Gottschalk composed the score for D. W. Griffith's *Broken Blossoms* in 1919. After a prologue by an onstage Chinese orchestra, the first cue came on the screen: "It is a tale of Temple Bells." A gong sounded and the story began, cue by cue.

Orchestra on Stage
f
And^te T It is a tale of Temple Bells
Largo T We may believe
Bells
p
Gong
mf
Trpts muted
1
there are
Oboe
p
Trem.
Fl. Ob
Cl. & Strgs
Trpts muted
Gong
Allegretto
Strings
p
Cello
Tymp
Strgs pizz
pp Harp
Ob. Cl.
pp
Strgs
Trpt.
Horns
Wood
Trpt
Strgs
Wood
p At the turnstile
pp Str. pizz
Ob. & Cl.
pp
Wood
Str. pizz
Wood
And^te 1 Street scene
Oboe Solo 2 Chinese girls
p
1
2 Full
poco rit
poco rit
Allegretto
Ob & Cl.
B'ss'n & Cello
Wood Horns
8va
Trgl

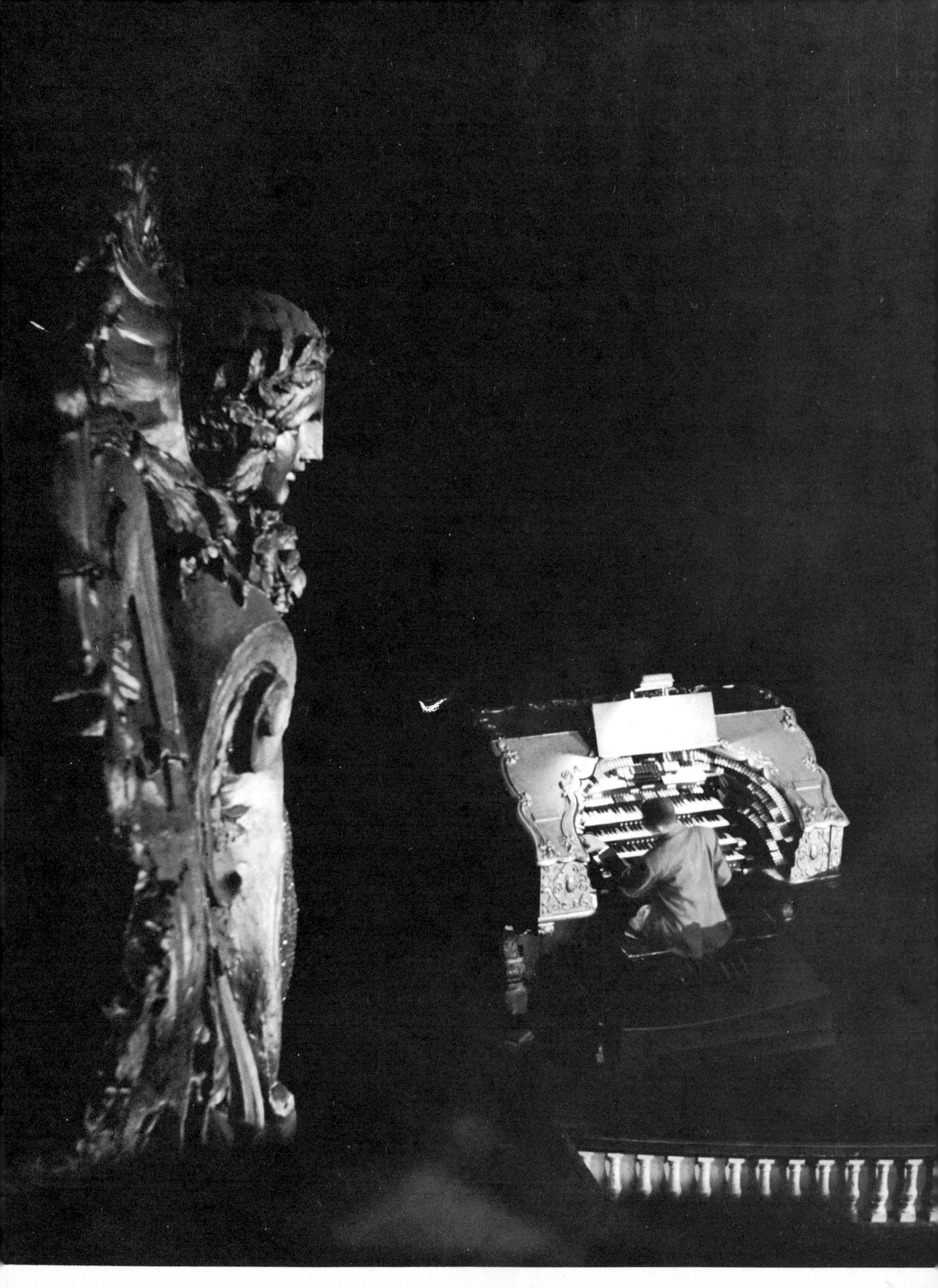

THE APOTHEOSIS OF THE MIGHTY WURLITZER

"If an exhibitor should be forced to choose between a poor orchestra and a good organist, he should consider it his duty to give the organist the preference. After all, it is quality and not quantity that really counts. Besides, it has been my experience that audiences would rather hear music played extremely well than extremely loud."

CARL EDOUARDE,
Musical Director of the
Mark Strand Theatre, New York—1921

Not many theatre managers could afford to keep a full symphony orchestra in the pit all day long, and, with the exception of the "super de luxe" houses—the Roxy, the Capitol, and the Paramount for example—certain shows on every day's schedule were designated as "accompanied by the grand organ." These were usually the first show in the morning and the last show in the afternoon; at the more popular "ladies' matinees," "supper shows," and "gala evening performances" audiences expected—and got—symphony with their Clara Bow.

The theatre organist did yeoman duty from the moment the house opened (usually at 11:30 for an audience of hooky-playing housewives and "sick" stenographers) until closing time (around midnight) when he provided an exit march for the bleary-eyed fans to stumble up the aisle to. After accompanying the first "non-de luxe" feature film, the organist had a few minutes for a cigarette while the orchestra played the overture for the first complete de luxe show. Then he had to get on his gilded behemoth and rise out of the orchestra pit "in a glow of light like a moon coming up over the sea," ready to put on his act—a ten- or twelve-minute stint that combined his talents for personality projection, console calesthenics, and choral direction.

It was in this capacity as flapper's darling, show-off and choirmaster that the theatre organist earned his weekly pay check. In the days before too many mediocre ex-pit-pianists-turned-organists had soured the public's taste for organ music (and the approaching "I-dare-you-to-entertain-me" attitude had begun to erode the "ain't-we-got-fun" spirit of moviegoers during the golden age) the brilliant and brilliantined young man at the Mighty Wurlitzer was one of the most popular fixtures on every program.

•

Few wonders of the movie palace brought more shivery pleasure to audiences (or caused more breast-beating among crusaders for Culture) than the Mighty Wurlitzer. Part one-man band, part symphony orchestra, part sound-effects department, the Wurlitzer was one of the most versatile instruments ever devised by man.

Of course, there were a score or more manufacturers of theatre organs, but the Wurlitzer basked in the same sweet sunlight of generic familiarity as the Frigidaire, the Victrola, and the Kodak. It might be a Kimball, a Robert Morton, a Möller, a Page, a Barton or a Marr & Colton (a few of the better-known makes), but to the average moviegoer, if it rose up out of the pit at intermission with a roar that made the marrow dance in one's bones, if rows of colored stop-tabs, lit by hidden lights, arched like a rainbow above the flawless dental work of the keyboards—if it could imitate anything from a brass band to a Ford horn to a choir of angels—gee, Dad, it was a Wurlitzer.

The Mighty Wurlitzer (and its counterparts) was as much a part of the movie palace as the electric lights that danced around the marquee, or the goldfish that swam in the lobby fountain. Inside the theatre the music seemed to bubble up and soar into the darkness of the balcony. Far below, bathed in a rose spotlight, was the organist perched in the maw of the great golden console. A flick of the finger, and chimes would call

The Mighty Wurlitzer in the New York Paramount Theatre is acknowledged the finest theatre organ ever built. Jesse Crawford designed it and it was installed in 1926 by Daniel Papp, the dedicated technician who gave it daily maintenance.

Ramona back beside the waterfall; a dramatic sweep of the hand and all would be silence save for the sobbing of the broken-hearted Tibia languishing in the left loft as it was comforted by its mate, the crooning Vox Humana over on the right—to the tune of "Prisoner of Love." A quick kick at the crescendo pedal, a lightning jab at the combination pistons, and the mood would change to joy again—all glockenspiels, trumpets, tubas, and snare drums—as an invisible MacNamara's Band marched across the balcony.

Then would come the community sing. The stage curtains would open to reveal a vine-covered cottage, seen faintly through a scrim drop, before which a tenor in blazer and white flannels stood poised ready to sing. On the scrim the lyrics of "My Blue Heaven" would be projected and the spotlight on the console would melt to a deep cobalt. In response to the organist's toothy cajoling, the balcony customers would be made to compete against the orchestra patrons with all the spirit of Yale versus Harvard. Then the ladies would be invited to sing a chorus while the men hummed; at ths point the organist would usually stop playing suddenly and, sure enough, the ladies *would* be singing and the men—somewhat sheepishly—humming. Finally, all "girls over twenty-nine" would be asked to stand up and sing a solo as the house lights blacked out, the organ opened up *sforzando*, and everybody nearly died laughing. Then it would be over. The organist, whirling around on his patented Howard seat, would beam and bow as he and his 18-carat leviathan would begin to sink inexorably into the depths. The golden spotlight would fade away, the music would grow softer, then nothing would be left but an ember glow from the pit and a complaining sigh from the organ chambers as the Vox Humana went to sleep.

Critics of the Wurlitzer called it blatant and vulgar, a threat to public morals. "Just how strong men become movie organists, whence recruited, and by what sinister inducements," wrote a detractor in 1928, "we have not been told. Perhaps, like the office of the French headsman, the calling is hereditary, taken up in earlier and more happy days when organs were strictly organs and not cluttered up with cymbals, sleigh bells and snare drums."

But the Mighty Wurlitzer could rise—hydraulically—above such sour-grape-eaters. It was, after all, capable of producing honest music of compelling emotional force when properly played and, in the hands of an expert, could work musical magic far beyond the limitations of any other single instrument. Without it the movie palace would have been as soulless as an armory.

The first organs in theatres were ponderous affairs, about as well suited to their purpose as a string quartet in a six-ring circus. All they had in common with the later Mighty Wurlitzer was the fact that they all depended on wind, keyboards, foot pedals, and pipes to make their music. And yet these early organs—nothing more than transplanted church instruments—brought a certain grandeur to the premises that set a movie *theatre* apart from the nickelodeon with its beleagured upright piano.

As movies grew longer and more ambitious, something more versatile than the piano was needed to cope with the flickering changes of mood and the demands for special sound effects that filled every scene in a photoplay. A full orchestra at every performance was out of the question for even the grandest of the early de luxe cinemas, and the organ was the ideal substitute. But a tubby church-type organ that made even "Waltz Me Around Again, Willie" sound like an offertory was woefully out of place, as audiences grew more discerning. Then, a few years before World War I, the answer to the problem came in the form of the Hope-Jones Unit Orchestra, built by the Rudolph Wurlitzer Company of North Tonawanda, New York — famous until then as importers of fine violins, makers of pianos and mechanical organs for merry-go-rounds.

Robert Hope-Jones, its inventor, a remarkable little Englishman with a shock of white hair that made him resemble Franz Liszt in a fright wig, had been tinkering with ideas for "liberating" the pipe organ ever since the turn of the century. A former telephone engineer, he pioneered the developments that were later to result in the theatre organ while modifying church instruments in England. The first of these innovations was electro-pneumatic action. This replaced the noisy and cumbersome "tracker" action (that literally tied the organ console to its pipes by hundreds of rods or stretched wires) with a system for opening and closing the valves in the pipes with electromagnets which were, in turn, controlled by sterling silver electrical contacts under each key on the console.

Operating on low voltage, this system made it possible for the console to be located almost any distance from the pipes (and in later days of splendor, on lifts that would raise and lower it in orchestra pits and, in many instances, revolve).

Hope-Jones also introduced stop tablets—tongue-shaped pieces of ivory in various colors which could be easily flicked by the organist's fingers to activate the organ's stops or voices. These tabs appeared in place of the awkward draw-knobs of the traditional organ. At the same time, he conceived the idea of arranging them in curved rows above the keyboards or manuals—thus the familiar "horseshoe" console was born.

But Robert Hope-Jones's most important contribution to the organ was his system of pipe unification. Up until that time all organs were built so that each manual could play only a limited number of the organ's basic sets of pipes or "ranks." The unit principle, by means of an intricate system of switches and relays, made it possible for every rank to be played from every manual, at many different octave pitches. Thus, a unit organ of six ranks (six separate sets of pipes each with a distinctive voicing) could be made to outshine in performance and tonal variety a "straight" organ requiring more than thirty ranks.

Hope-Jones often made the comparison between the Unit Orchestra organist and the painter who had six fundamental colors to work with: "By mixing these six colors, he can get a limitless number of various shades, because he can mix them at will. With a 'straight' organ of six ranks, one is very limited in musical results, whereas with a Unit Orchestra of six ranks, one has a really remarkable number of possible combinations."

Robert Hope-Jones, 1859-1914, the eccentric genius of the Unit Orchestra, father of the Mighty Wurlitzer.

Rudolph Wurlitzer, 1831-1914, who gave the Unit Orchestra his name and set it on the road to glory.

By devising ways to combine ranks into new voices or stops, and by developing many completely new solo voices, (his Diaphone was invented first as a fog horn, and is still used by the Coast Guard as such) Hope-Jones enriched the organ, made it an instrument of incredible flexibility, and opened up a dazzling new career for the stuffy church mouse in the outside world.

After coming to the United States, the brilliant if more than slightly eccentric inventor built a number of instruments for churches, hotels, and small theatres before going bankrupt. When Andrew Carnegie was introduced to Hope-Jones in 1908, he remarked: "I want to have an organ overwhelm me with the feeling of how miserable a sinner I am," and the inventor obliged by giving him a demonstration on the giant organ his company had just built in the Tabernacle at Ocean Grove on the nearby New Jersey shore. Unfortunately, Carnegie was not made to feel sufficiently miserable to invest in the organ works in Elmira, New York, and in 1910 the Hope-Jones Organ Company (motto: "*Scienta Artem Adjuvat*") was absorbed, along with its owner and all his patents, by the Wurlitzer Company.

Quick to realize the exciting possibilities of the instrument as a theatre organ, Wurlitzer launched into full-scale production of the Wurlitzer Hope-Jones Unit Orchestra. By the time the first real movie palaces were being built, it was ready to roar to fame and glory. Its success, however, came too late to have meaning for its inventor. With less and less to say about the destiny of his brain child, he became increasingly bitter. "The Wurlitzers say that they are going to build better organs without me," he wrote to a friend on April 1, 1914. "I am no longer at the factory, neither have I anything to do with the jobs outside. I am a 'gentleman at large' on $60 per week. They refuse to stop the $60 because it would give me my freedom."

On September 13th of that year he perfected his last invention—a new and bizarre way of committing suicide. He attached a rubber tube to a gas jet and fitted it with a T outlet. One end was firmly taped with adhesive to his mouth which was sealed closed, as was his nose. Gas escaping from the other end of the T outlet was ignited so that there would be no danger to others from escaping gas after his death.

•

"Unit Orchestra" was the perfect name for the new instrument. Basically a pipe organ, its voices were designed to imitate all the instruments of the orchestra, plus many that no orchestra (or organ) had ever heard before. These musical marvels were equipped — in addition to their thousands of pipes—with a battery of bass drums, snare drums, xylophones, glockenspiels, marimbas, grand pianos (that could also imitate banjos and mandolins), sets of tuned sleigh bells, chimes, triangles, cymbals, castanets, Chinese blocks, tambourines, tom-toms, gongs, and saucer bells.

But that wasn't all. The organist at the console of the Unit Orchestra could conjure up nightin-

ORGAN SUPPLY CORP.
540-550 East 2nd Street
ERIE, PA.

MANUFACTURES

CONSOLES ACTIONS PIPES SWELL ENGINES
CHESTS TREMOLOS WOOD PARTS SUPPLIES, ETC.

This "toy counter" could be hooked on any organ (the name brands, of course, had their own). It included (l. to r.) snare drum, bird call, tambourine, auto horn, wood block, steamboat whistle, triangle, castanets, surf effect, tom-tom, crash cymbal and bass drum — a real one-man band.

gales, canaries, galloping horses, steamboat and train whistles, auto horns (both "honk-honk" and "ah-*oo*-gah"), fire-engine sirens, airplanes, hurricanes, swishing surf, rain on the roof, telephone bells, door bells, trolley bells, and the sound of smashing crockery—all with hair-raising verisimilitude. There was virtually no mood or situation on the silent screen that a quick-thinking and agile organist couldn't heighten with some musical theme and mechanical effect from the Wurlitzer Hope-Jones Unit Orchestra's bag of tricks.

As Mighty Wurlitzers began to replace the galumphing old church organs in theatres, a whole new breed of organists appeared on the scene. No longer was it enough to be able to play "Kammenoi-Ostrow" and gems from Tchaikowsky; you had to know your Irving Berlin, and you had to know how to hit the Jazz Whistle, the Chinese Block, and the Bass Drum buttons in proper sequence when Fatty did a pratfall up on the screen. But, above all, you had to be an entertainer, capable of doing an act on your own as part of the ever-more-elaborate presentations that were becoming the vogue in the movie palaces.

One name towers over all the rest: Jesse Crawford. This is not to say, of course, that there weren't dozens of theatre organ virtuosi who were real stars in the entertainment world of the Twenties: Eddie Dunstedter, Lew White, Stuart Barrie, Irma Glenn, C. A. J. Parmentier, Ann Leaf, Henry Murtagh, Don Baker, Milton Charles, Emil Velazco, Milton Slosser, Sigmund Krumgold and Dick Leibert were only a few, and every city had its local favorites. But Jesse Crawford, presiding at the great cream and gold console of the Wurlitzer in the Paramount Theatre in New York, was king.

"The Poet of the Organ," as he was billed, got his musical start playing the cornet at the age of nine in an orphanage band. At sixteen he was banging the piano in a Spokane nickelodeon. Then he got a job as a full-fledged theatre organist playing on a non-unified Skinner organ in Seattle's de luxe (10¢) Alaska Theatre. Shrewd Sid Grauman rescued the young "poet" and sat him before his first Wurlitzer in Grauman's new Million Dollar Theatre in Los Angeles in 1917. Then Balaban & Katz stole Crawford away from Grauman in 1921 to open their super-de luxe Tivoli Theatre, where he played until the Chicago Theatre opened the following year.

In a nearby Loop theatre, the Roosevelt, Crawford heard a dark-eyed beauty named Helen Anderson playing the Roosevelt's Kimball organ. Soon they were married and the Chicago Theatre Wurlitzer sprouted another console, with Helen joining Jesse in the duets that made them one of the best-paid "Mr. and Mrs." teams in show business.

By 1926 the Crawfords were lured away from Chicago to New York to open Adolph Zukor's huge Paramount Theatre at the "Crossroads of

The Lunt and Fontanne of the Mighty Wurlitzer, Mr. and Mrs. Crawford were a twin-console attraction at the Chicago Theatre and the New York Paramount, and later toured the U.S.

the World," Times Square. As part of the arrangement, Jesse was given *carte blanche* in designing the Wurlitzer for the new theatre. A dream organ in every sense, it was the biggest Wurlitzer yet built—thirty-six ranks of pipes plus a complete "hardware" department, including a set of tuned timpani. Appropriately, Crawford's inaugural program on the new organ, the night the theatre opened, was "Organs I Have Played," and it must have taken considerable ingenuity to make the "Queen Mother of all the Wurlitzers" sound like the little Skinner groan box in Seattle that had set the poet of the organ on his way to fame.

Crawford's national celebrity was due not so much to his performances in Chicago and New York theatres as to the more than a hundred Victor records that he turned out between 1925 and 1935. It was an austere Victrola or Panatrope indeed that did not house one or more records of Crawford's exquisitely polished "gems of miniaturization" on the Wurlitzer. Accordingly, when he made the move from Chicago, where he had been recording on a small instrument in the Wurlitzer showroom on Wabash Avenue, to New York, where he first recorded on another showroom organ in the Wurlitzer store on Forty-second Street, he felt the need for a larger and more versatile organ to show off on. Victor's carbon microphones were not up to capturing the sound and acoustical setting of the big Wurlitzer in the Paramount Theatre; one rumbling note from the towering thirty-two-foot Diaphone pipe would send the recording stylus screeching right off the wax. So another Wurlitzer was designed by Crawford expressly for recording, an instrument of twenty-one ranks, and this was installed in a studio in the Paramount building two years later. Here he (and many other organists) made records and supplied much of the pipe-organ music that used to saturate the networks back in the tenderly recalled days of live radio.

Newspaper ads for the Paramount always featured the Mighty Wurlitzer and its mustached little master. "Hear Jesse Crawford at THAT organ!" was a popular line. Crawford's forte was the ballad and the descriptive novelty; Mrs. Crawford went in for the more rhythmical sort of thing. The first time they appeared in tandem at the Paramount, *Variety* covered the event in detail, and reported that after Crawford had finished playing "Tonight You Belong to Me" and "Moonlight on the Ganges," an announcement came on the screen stating that there had been many requests for jazz music on the organ and that the greatest jazz organist in the world was about to be introduced with great pride on the part of Jesse, it being added that he ought to be proud; it was his wife.

"Mrs. Crawford took her place at the great console," wrote *Variety*, "and started with 'Black Bottom' which always brings applause, and followed with 'This Is My Lucky Day,' and finally 'Blue Skies.' For the final chorus of the last number, Jesse unfolded a small secondary console at the side of the organ and joined his wife, with the double effect being nothing short of remarkable. This concert, as presented, is strong enough to headline any vaudeville bill, no matter how big it is if there is an organ in the house. Let's have more of Mr. and Mrs. Crawford."

Soon there was no need for Jesse to play on the little portable keyboard that had been attached to the console when Helen took over for her jazz interludes on the Wurlitzer. In a few months the Paramount installed a second console in the orchestra pit, a twin to the original one, plus two more consoles that could be rolled out on the stage, and the Crawfords were playing organs all over the place. Helen, incidentally, garnered great publicity from the fact that she was the only woman in show business who had her gowns made backward—all the decoration was on the rear, because that was what audiences saw the most of.

When John Philip Sousa's Band played a week at the Paramount, trained nurses were stationed in the aisles "to assist those overcome by the sheer magnitude of sound when the Sousa Band, the Paramount Grand Orchestra, and Mr. and Mrs. Jesse Crawford at the twin consoles of the Mighty Wurlitzer all joined together in their rendition of 'The Stars and Stripes Forever.' "

•

Early in 1926 Roxy had read of the plans for the Wurlitzer organ that was being built for the Paramount Theatre then under construction. The shrewd showman realized that this organ, with Jesse Crawford at the throttle, would be one of the brightest stars in Adolph Zukor's galaxy, and he decided that when his own Roxy Theatre opened the next year, it must "out organ" the Paramount. Roxy had already contracted with the W. W. Kimball Company of Chicago to build three organs for his new theatre—a large instrument in the auditorium, and two smaller ones in the broadcasting studio and the Rotunda respectively. Perhaps he would have preferred Wurlitzers—but Roxy had to be different, so Kimballs they were. In any event, instructions went forth to Kimball to build the most flamboyantly "visible" organ in the world for the new Roxy Theatre. Kimball rose to the occasion (and to the juiciest single order any theatre organ builder ever got) and promised Roxy that he would have something that would "knock 'em dead" on opening night.

The something was a mammoth antique-gold console with five manuals, flanked by two equally aureate consoles of three manuals each, all rising out of the pit on separate worm-gear lifts. Three consoles with three organists working away simultaneously (Roxy issued an edict that he "wanted to see plenty of feet moving all the time") was a

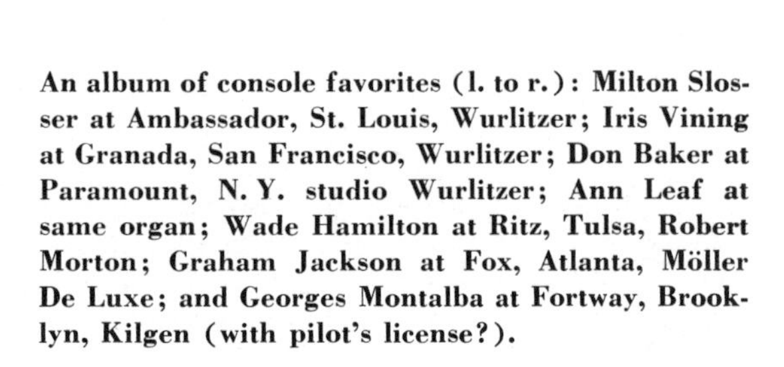

An album of console favorites (l. to r.): Milton Slosser at Ambassador, St. Louis, Wurlitzer; Iris Vining at Granada, San Francisco, Wurlitzer; Don Baker at Paramount, N. Y. studio Wurlitzer; Ann Leaf at same organ; Wade Hamilton at Ritz, Tulsa, Robert Morton; Graham Jackson at Fox, Atlanta, Möller De Luxe; and Georges Montalba at Fortway, Brooklyn, Kilgen (with pilot's license?).

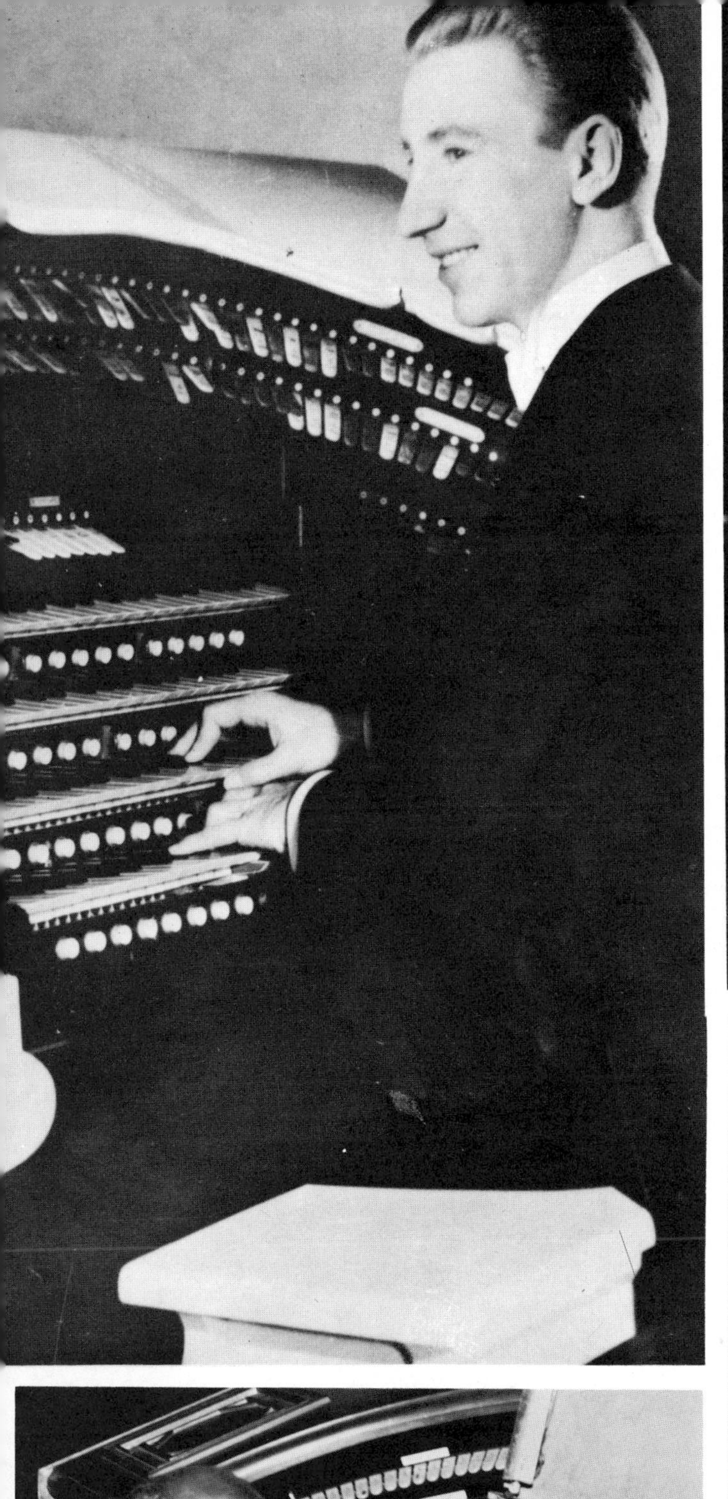

sight calculated to make Roxy audiences and the powers at the Paramount sit up and take notice.

C. A. J. Parmentier (who, with Dezso Von D'-Antalffy and Emil Velazco, opened the Roxy organ) recalls that when he first hove into view on opening night aboard the five-manual main console in the center of the orchestra pit, there was a sharp intake of breath all over the theatre, followed by applause which grew as D'Antalffy and Velazco arose in turn at the neighboring three-manual consoles.

But, so far as the organ itself was concerned, it was a case of three tails wagging a dog. The Kimball organ in the Roxy had only twenty-nine ranks, as compared to the Paramount Wurlitzer's thirty-six, and it suffered from being installed in chambers under the stage, instead of in the orthodox arrangement on either side of the proscenium. Roxy's idea in putting it there was a fairly logical one: he wanted the organ tone to come from the same general direction as the orchestra. But when the huge orchestra platform, supporting Erno Rapee and the 110-piece Roxy Symphony, was elevated to "overture" level, the organ pipes—which spoke through grilles at the bottom of the orchestra pit—were whistling in Dixie.

The Roxy organ had one division, however, which was not smothered under the stage. The Fanfare Organ, consisting of three ranks of pipes called the Military Bugle, the Fanfare Trumpet, and the Fife, was housed in a small chamber high above the stage to the right. When, as the Pilgrims' Chorus from *Tannhäuser* reached its final crescendo on opening night, and the Fanfare Organ joined the fray from on high, loud and clear, the effect was electric. After the organ concert was over and the three organists had descended back into the pit, each received a sweaty bear hug from the suspendered and perspiring Roxy.

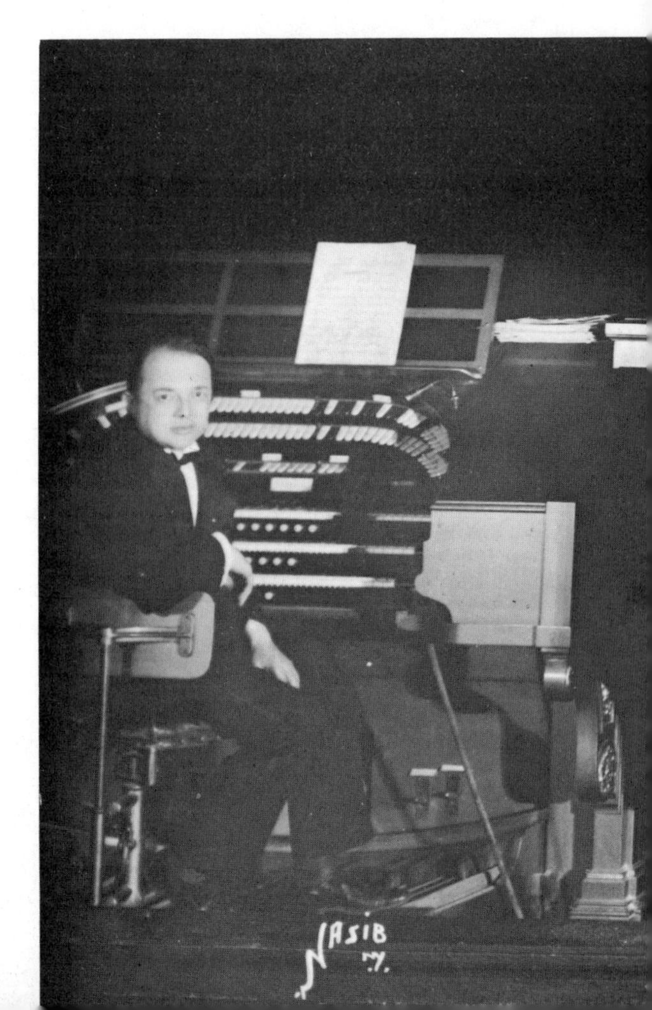

Lew White, who had played in the Rotunda on opening night but soon promoted himself to the chief organist's job in the Cathedral, tried to duplicate Jesse Crawford's success on records. He recorded on the Kimball organ in the broadcasting

The Roxy organs on parade. The Kimball Company displays its handiwork in Chicago factory (above, l. to r.): console for studio organ; brass, main and woodwind consoles for theatre organ; duplex player console for Rotunda organ.

The world's first pipe organ trio (below): Dezso Von D'Antalffy, Lew White and Dr. C. A. J. Parmentier. April 30, 1927.

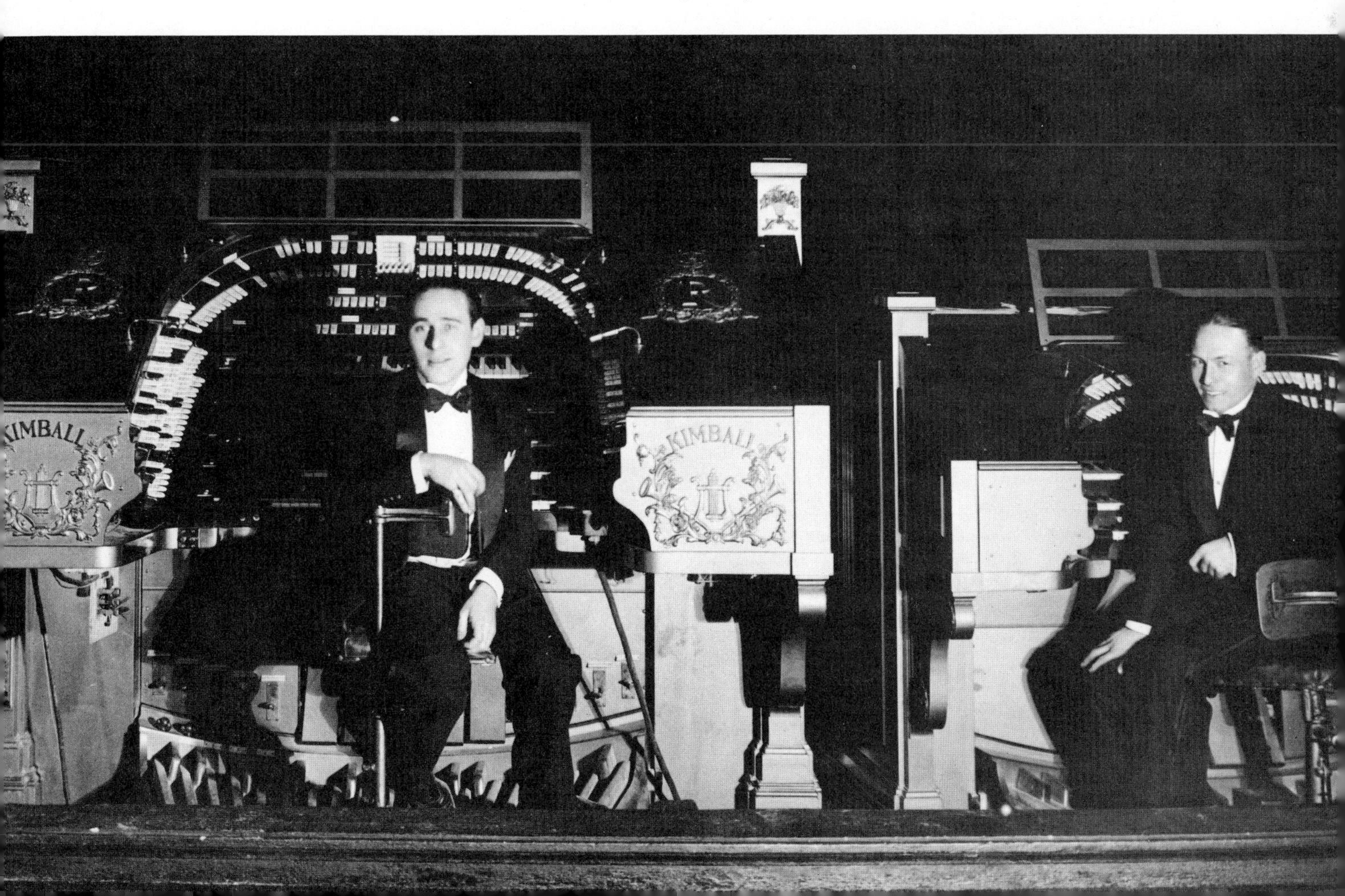

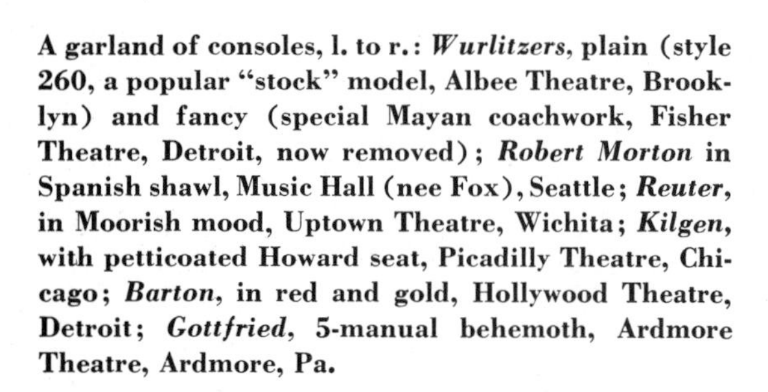

A garland of consoles, l. to r.: *Wurlitzers*, plain (style 260, a popular "stock" model, Albee Theatre, Brooklyn) and fancy (special Mayan coachwork, Fisher Theatre, Detroit, now removed); *Robert Morton* in Spanish shawl, Music Hall (nee Fox), Seattle; *Reuter*, in Moorish mood, Uptown Theatre, Wichita; *Kilgen*, with petticoated Howard seat, Picadilly Theatre, Chicago; *Barton*, in red and gold, Hollywood Theatre, Detroit; *Gottfried*, 5-manual behemoth, Ardmore Theatre, Ardmore, Pa.

studio, whence Roxy's Gang hit the air waves every Sunday evening. There was an exclusive arrangement with Brunswick Records that would credit both the Kimball Unit Organ as well as the Roxy Theatre on record labels (the Paramount had never been included in the Crawford records' "Played on the Wurlitzer Organ" slogan). But the nine-rank studio organ at the Roxy was no match for the twenty-one-ranker in the Paramount's studio, and White's recordings had a "goaty" tone about them that made their popularity run a poor second to Crawford's.

Down in Atlanta, Georgia, an organ was installed that had the distinction, for three years, of being the largest organ in any theatre in the world. The Möller De luxe instrument in the Fox Theatre there had forty-two ranks of pipes—six more than the New York Paramount Wurlitzer—and was surpassed in number of ranks (though not in sheer manic versatility) only when the Radio City Music Hall opened in 1932. The Music Hall's twin-consoled Wurlitzer boasted fifty-eight sets, and has never been outranked.

Fifteen ranks alone on the Atlanta Fox organ were given over to string tones (Gamba, Cello, Viola, String Bass, Salicional, Bass Viola, Orchestral Violin, Solo Violin, and so forth) which were produced by organ pipes but which sounded surprisingly like the real thing. Among its more novel accouterments were "trick coupler" tabs which, when activated, would make a single note struck on a key sound like a full chord. It also possessed a row of "effects" tabs above the manuals that presented such startling possibilities as "Ding Dong I" and "Ding Dong II," "Slap Sticks," "Storm," "Crash" and "Grand Crash." These last two effects were produced by a large steel mesh box mounted on an axle in one of the organ chambers and filled with nuts and bolts, scraps of crockery, and pieces of tin. When a crash was needed, the box made a quarter-turn on its axle; when the situation called for a grand crash, the whole business started revolving until the organist released the tab. There was no "Panic" button, however.

The organ in the Atlanta Fox may not have been equipped for panic, but there was one make of theatre instrument that came close. This was the "Symphonic Registrator" organ built by Marr & Colton. Above the usual row of stop tabs controlling the standard ranks of the organ (Tibia, Vox Humana, Flute, Strings, etc.) there was an additional row of tabs. Each of these was marked by a spot of color—violet, blue, green, red, pink, yellow, orange—and these were inscribed with a catalogue of emotions and situations that would make the most brazen scenario writer blush for shame: "Love (Mother);" "Love (Passion);" "Love (Romantic);" "Quietitude;" "Jealousy" (green spot); "Suspense" (blue); "Happiness;" "Hate;" "Mysterious" (gray); "Gruesome" (black); "Pathetic;" "Riot" (red)—to list only a few. Each of these tabs controlled preselected combinations of ranks which produced tones suitable to the indicated mood—it was up to the organist to supply the melody. When a burning building appeared on the silent screen, the organist had only to flick the red-spotted Fire tab and a siren and bells and lots of trumpets sprang into action in the organ chamber; when Vilma Banky slithered over the incorruptible Ronald Colman the organist was ready with Love (Passion) and Agitation. For the inevitable chase scene there was, of course, "Chase," which turned out to be "Fire" minus the siren and bells.

While nearly every theatre organ had one or more ranks of pipes which imitated stringed instruments, it remained for the Robert Morton Company to come up with the real thing. This musical curiosity was known as the "V'Oleon" and consisted of a lyrelike instrument with sixty-one stout steel strings. When brought into play, resin-coated rollers spun against the strings to produce

The Marr & Colton Symphonic Registrator (top) provided the silent movie organist with a catalogue of built-in emotions. Top row of tabs begins: "Love (Mother)," "Love (Romantic)," "Love (Passion)." Biggest Marr & Colton ever built (bottom) was 5-manual, 24-rank organ for Loew's Theatre, Rochester, N. Y. Buttons under manuals, found on all organs, control pre-set combination of ranks.

ACCOMPANIMENT
SOLO
PEDAL
PEDAL
ACCOMP
1 2 3 4 5
1 2 3 4 5
SOLO
GREAT
CRESC
SOLO
GREAT
ACCOMP
ORCH

a tone not unlike a cello, the whole thing being played from the console.

Before 1930, when the advent of talkies called an abrupt halt to the installation of any more organs in theatres, nearly every movie house of more than three-hundred seats had some kind of an organ in it. The smaller theatres were equipped with either little two-manual instruments of seven or eight basic ranks, or with devices resembling player pianos bolstered with pipes and an array of special effects. These remarkable instruments were housed in large cabinets, flanking the piano case, and usually were installed in what would have been the orchestra pit in a larger house.

The Robert Morton Fotoplayer was a typical "pit organ" (although nearly all the organ builders made a similar economy model). The Fotoplayer was distinguished by the fact that its effects (gong, drum, whistle, klaxon, etc.) were operated by a row of ropes with wooden handles that dangled in front of the operator. It required the combined talents of a streetcar motorman and a carilloneur to keep everything going at once. In some of the more modest cameos around the country there was no regular musician on the payroll; instead, a small boy would be put in charge of the Fotoplayer (which could be operated by standard pianola rolls as well as by hand). All he had to do was rewind the roll, or occasionally change it when the audience became restless, and yank the daylights out of the sound effects whenever he felt the action on the screen dictated—which was always.

In the movie palaces, of course, things were quite different. The mightiness of the Wurlitzer within came to be a status symbol as newer and grander theatres were opened. Consoles were designed to match any decor—an oriental temple with jewel-eyed Buddhas guarding the rest rooms might boast an organ console that rose from the depths crusted with crimson dragons and lacquered lotus blossoms. A vast Italian garden, open to an Eberson sky, could feature a console frescoed with peacocks or (as in the case of the Paradise Theatre in Chicago) acrawl with alabastine cupids.

In Denver, the Isis Theatre reached new meteorological heights when a lightning machine was installed as part of the organ. The organist, with the help of hundreds of light bulbs (and some pneumatic motors which operated quick-acting switches), mounted behind the organ grilles over the proscenium could set off an electrical storm during the course of the *William Tell Overture* that had the audience galvanized in its seats.

•

"What, you must often wonder," mused a critic called "Screencomber" in the late Twenties, "happens to the organist when once he sinks down into the earth's bowels? To what red hell . . . ? Exactly! Simply that he returns to the Organists' Purgatory, a vast subterranean limbo populated by lost souls and consoles of organists, doomed to wander forever in this kingdom of the damned while the Mightiest Wurlitzer of Them All plays a perpetual interlude, alternating between *Poet and Peasant* and *William Tell*, except on feast days when they substitute *The Whistler and His Dog*."

"Screencomber" had it all wrong. The organist spent the rest of his nights up in the organ chambers tuning pipes and patching leaking wind chests so that on the morrow he and his beloved might rise again in golden and pitch-perfect glory, while the applause swirled around them and all the girls —over twenty-nine, under twenty-nine, who cared? —threw back their boyish bobs in a love song. That's why theatre organists were born.

Some units of a Unit Orchestra. This view of the right half of a Wurlitzer organ chamber shows parts of some of the more typical ranks. From bottom to top: Brass Saxophone, Brass Trumpet, Quintadena, Oboe Horn, Krumet, Tibia, Salicional, Orchestra Oboe and Kinura. The Tibia (square wooden pipes with stoppered tops) is the sobbing tone most characteristic of theatre organ; the Kinura (top row with flared mouths) is reed tone, sharp and bleating; Salicional (slender pipes behind Tibia) is violin-like string tone; Brass Trumpet (taller flared pipes behind Sax) looks and sounds like a — brass trumpet.

. . . AND ON THE GREAT STAGE

"The major difference between a de luxe and any other movie show consists of closing the curtains after the feature film, re-opening them for the shorts, closing them again and re-opening them for the news."

—ARTHUR MAYER
in MERELY COLOSSAL — 1953

Remember how the curtains used to start closing as the picture unreeled the final embrace, so that "The End" was projected on the rippling velour? And how the colored spotlights from the sides of the proscenium used to play across the curtains as the footlights faded up from purple to violet to red to orange and their final golden burst, making the fringe and the rhinestone butterflies sparkle? This spelled out real *class*, and when you saw it, you knew now something marvelous was about to take place.

Every self-respecting presentation house had numerous sets of curtains. They were numbered, from the proscenium arch back; the phrase "in one" meant, in movie-palace and vaudeville parlance, that an act took place before the No. 1 curtain. But before the numbered curtains came, first of all, the grand drapery, festooned in swags and furbelows across the top and sides of the proscenium. Directly behind it hung the asbestos curtain (hand-painted with a scene following the motif of the auditorium) that was lowered and raised once a day, according to law, to assure its working properly in case of a backstage fire. The house curtains (and these were usually given the No. 1 designation) came next—elaborate affairs, heavily fringed and betasseled and sometimes ornamented with sequin peacocks or huge jeweled flowers. Behind these came many sets of "travelers" (curtains that closed and opened from the sides in the standard way). No. 2 traveler usually was the one that drifted behind "The End," and was used between the various film portions of the program.

The screen (or picture sheet, to use proper theatrical terminology) was usually of the Magnascope variety, which meant that the black velvet masking on the top and sides could be opened up to enlarge the screen opening whenever Fox Grandeur features, or certain scenes in other films, were shown that required a larger screen than the standard twelve by sixteen feet or (in the largest theatre) fifteen by twenty. Frequently the newsreel was shown on the Magnascope screen, or the lyrics to the organ sing-alongs would be projected on it, surrounded by colored pictorial effects. The whole affair could be "flown" (raised up out of sight) when the stage show started, and frequently would be brought back down for a comedy or Travelogue or newsreel at various points in the presentation before being anchored for an hour or so to show the feature.

Hanging behind the picture sheet were numerous travelers in various colors (always including one set of black velvet) and materials—plush, silk, velvet, satin—and often one or two of these could be operated as "tabs" (curtains that loop back up opera-house style). There were also a number of drop curtains, including one of scrim (cheesecloth that was opaque as long as all the light hit it from in front, but grew gradually transparent as the lights were brought up from behind). Some theatres had curtains of "dream cloth"—a sheer fabric interwoven with metallic threads that caught the light in a lovely way. And sometimes a particularly elaborate and versatile drop curtain would also work as a traveler *and* a tab.

At the back of the stage was the cyclorama, a plain canvas drop painted a rich blue and slightly curved; when properly lit this would simulate an outdoors sky. Often clouds were projected on it and sometimes constellations of little electric lights were fixed permanently over it so that night scenes could be created. Along the sides of the stage—the wings—hung vertical sections of curtain that masked off the backstage from the audience and still allowed dancers and actors to enter and exit, and corresponding strips of curtain hung from side to side across the top of the stage about eight feet apart called "borders." All these curtains were usually standard equipment in de luxe theatres; in addition to these, every traveling unit show brought along its own special curtains for every act, and these had to be hung,

Projects Everything but the Motion Picture

along with tons of scenery, each week before each new show opened.

The lighting of a de luxe house was one of its real glories. At the back of the house, in the projection booth, were usually five or six spotlights aimed at the stage through a wide opening. These could be swung around or tilted to follow any act, and had iris diaphragms set between the lenses so that the size of the spot could be made as big as the stage opening itself or small enough to pick out a single performer's head. In front of the lenses were racks to hold various colored gelatin slides; illumination for these spotlights was by powerful 1,000-candle arc lights, similar to those used in the movie projectors.

Along the edge of the balcony were usually a row of spotlights and floodlights mounted in concealed boxes. These had motor-controlled color wheels mounted on them so that the electrician could make them change colors untouched by hand. In the largest theatres, except in atmospherics where it was obviously impossible, there was another spotlight booth (this one manned) high in the dome of the ceiling to beam lights down on the orchestra pit or on actors performing on the stage apron.

Over the proscenium arch were more concealed (and fixed) spotlights beamed down on the orchestra pit, and these also were equipped with remote color-changing devices. Other things happened here, too. Once, during a gala Christmas week in the Fox Theatre in Atlanta, it snowed on Enrico Leide and the Fox orchestra as they played a symphonic arrangement of "Jingle Bells." The musicians spent their time between shows sifting the bleached corn flakes out of their instruments.

The projection booth was also equipped with a Brenograph, a super magic lantern that not only projected song slides for the organ interludes, but an endless variety of scenic effects by means of multiple lenses and moving slides and

intricate fades and dissolves. Some of the effects listed in the Brenkert Company's (manufacturer of this 100-volt Aladdin's lamp) catalogue for 1928 were Aurora Borealis, Babbling Brook, Blizzard, Descending Clouds (for imaginary ascension trip), Flying Angels, Flying Birds, Flying Butterflies, Fire and Smoke, Flames, Lightning, Fast-moving Dark Storm Clouds, Slow-moving Fleecy Clouds, Moving River, Ocean Waves, Rain, Sand Storm, Snow, Volcano in Eruption (with flowing lava and rain of fire and ashes), Waterfall, Waving American Flag, Flying Fairies, Flying Aeroplanes, Falling Roses, Twinkling Stars, Rainbow, and Rising Bubbles. So equipped, the operator had at his finger tips the power to conjure up disaster, holocaust, and a plague of butterflies, to say nothing of the capacity to launch everything from a steamship to an aeroplane to Little Eva (complete with angel escort).

Footlights usually consisted of four circuits of lights, each circuit being a row of bulbs in reflectors with glass color filters in front—red, blue, green and white, from which all colors of the spectrum could be produced. Each circuit (and this was true of every circuit of lights in the entire house) was on dimmer controls so that any desired combination of colors and shades could be mixed at the switchboard. Additional banks of similar lights were mounted vertically just inside the proscenium arch, while others, called borders, hung along the top of the stage paralleling the footlights. Backstage were dozens of floodlights and spotlights, each provided with mounts for gelatin slides. Gelatin was used because it was less flammable than celluloid or plastic, less breakable than glass, and much cheaper to replace. It came in seventy-five standard colors, ranging from No. 0 (clear) through Medium Pink (popular No. 4), Dark Rose Purple (No. 15), Medium Blue Special (No. 32), Moonlight Blue (No. 41), Light Lemon (No. 50), Dark Straw (No. 56), Amber (No. 59), Fire Red (No. 67), to Special Chocolate (No. 75). You could have your choice of flavors.

Light bridges were constructed backstage so that spotlights could be projected at angles from the wings and from overhead as well. The cyclorama had special fixed floodlights to bring it to life, and in at least one theatre (the Roxy) a special curved row of footlights was installed in the stage floor around the base of the cyclorama that would tilt open, or close down to form part of the stage floor when not in use. In theatres where the orchestra pit boasted a "band car" (a platform on wheels that would propel the orchestra off the lift and back onto the stage proper) the footlights were of the disappearing variety. (See page 212.)

The orchestra-pit elevator was either a hydraulic or a worm-gear driven affair that could raise an entire symphony from sub-basement to stage level in half a minute. The elevator platform was entered by small doors from the area under the stage; the elevator mechanism had a safety feature that prevented it from moving until all the doors were firmly locked. Then the conductor, on signal from the stage manager, would press a button and up they would come. Another button, marked "Overture," would automatically halt this musical juggernaut at a level about three feet below the footlights. A second button, marked

Backstage at the Roxy was a confusion of lights, wires, curtains and ropes (above). Bank of floor lights for cyclorama are seen in extreme right; a scrim drop is flown above; sections in floor are elevator platforms. Roxy switchboard (upper left) was fearsomely complex; wheel at center is master dimmer control.

The silhouette spotlight (above) pops out of the Roxy stage by remote control (see page 234).

"Picture," would lower the whole affair so that the harp and the bass fiddles didn't ruin anyone's view of Norma Talmadge while they played the accompaniment to *Woman Disputed* (with its hauntingly emetic theme song, "Woman Disputed, I Love You"). A third button would raise the orchestra lift to full stage level (if they were on a bandwagon) and another would set them rolling back onto the stage.

The organ console was mounted on a similar elevator and the organist had the same choice of levels at which to perform. Many of the orchestra elevators had, in their center, still another elevator on which a grand piano was usually mounted. This, too, could ascend independently, though it was a bold pianist who would risk the perils of solo flight up to the footlights with nothing but twenty feet of open orchestra pit surrounding him and his Steinway on three sides. Sometimes this piano lift was used to deliver a name act or a star attraction to the stage in a spectacular manner; Texas Guinan rose into view this way once at the Capitol in New York and remarked at the end of her ascent, "Well, you can't keep a good girl down."

The stage itself was usually divided into at least two elevator sections. This enabled thirty-six chorus girls to plunge out of sight kicking, and emerge a moment later after a quick change of costume, on another elevator without missing a beat. Or multilevel sets could be quickly achieved without laborious carpentry work, one of the favorites being the "cabaret set" in which the orchestra was on one level, high at the back of the stage, groups of actors at tables were on another level in front of them, while adagio dancers and comedians performed on the front part of the stage itself.

The switchboards controlling these miracles of lighting and levitation were beyond imagination in complexity. It is sufficient to say that every light of every color in the theatre (except those operated from the projectionist's booth) could be turned on or off or dimmed or faded singly and in combination with any number of other lights, as switches were thrown, levers pulled and wheels turned. All the rising and falling orchestras, organ consoles, pianos, red-hot mamas, chorus girls, and variety acts could also be set in motion from the theatre's nerve center, the switchboard.

Keeping a firm hand on all that happened backstage was the stage manager. A theatre's staff might boast a producer, director, ballet master, scenic designer, and conductor, each with varyingly volatile temperaments, but when the show was on, the stage manager was the undisputed boss. He operated from a little lectern near the switchboard at stage right, (or "prompter's" side; the left side of the stage was always known as the "O.P." — or "opposite prompter" — side coordinating the confusion of lights, curtains, elevators, music and tap shoes with his cue sheets. He kept in constant touch with his lieutenants by means of the house phone or annunciator system, by buzzers, and by wig-wags to the electricians and scene shifters. In some of the more pretentious theatres he was such an important personage that he even dressed the part. Bill Stern, who was stage manager at the Radio City Music Hall during that theatre's first rocky year before he turned to sportscasting, recalls that he always appeared in black tie and dinner jacket for the evening shows, though nobody ever saw him except the performers.

Loew's Kings Theatre in Brooklyn (above), Rapp & Rapp, architects, shows off some of its mechanical wonders on day before opening. In the pit the organ console is at stage level; orchestra lift is at "picture" position, while piano lift is at "overture" level. Stage elevator "B" is raised to full height (top of staircases) while "A" is up 2 feet for stage bandstand. A set of pre-focused spotlights is being flown to position on a batten; Vitaphone horn is moving off to right on its track. Note truly grand drapery; behind it the teaser, masking stage opening to rectangular shape. The Kings (named after Brooklyn's county) was opened in 1929 as one of the five Loew's "Wonder Theatres" (others: Valencia in Jamaica, Long Island; Paradise in The Bronx, Jersey, Jersey City, 175th Street in Manhattan. All had identical twenty-three rank Robert Morton organs, **but only the 175th St. organ is playable; the other consoles (some of which were long buried under concrete slabs since the halcyon days of the early fifties), were eventually removed.**

"SHOW US THE REALM WHERE FANTASY REIGNS"

"Although movie audiences were supposed to be composed entirely of incurable lowbrows, Roxy gave them highbrow entertainment and made them like it. He gave them grand opera one week and the next week beautiful orchestrations of the collected works of Irving Berlin; his ballets ranged from 'Les Sylphides' to the Charleston; he presented tabloid versions of the Gilbert and Sullivan operettas, beautifully staged and sung; he offered violin solos by the great Frederic Fradkin and saxophone solos by Rudy Wiedoft."

—ROBERT E. SHERWOOD, 1927

In the Twenties, while Roxy was elevating public tastes in New York, other equally creative forces were at work in different ways in other parts of the country. The Rothafel idea was to present a varied program of music and spectacle—a Spanish dance, a Russian chorus, a French ballet, a patriotic pageant, all on one bill. In Los Angeles, Sid Grauman, on the other hand, conceived his shows as prologues to the feature film, with each act and setting relating to the theme of the picture it preceded.

Other stage-show producers, notably John Murray Anderson at the New York Paramount, believed in presenting miniature musical comedies or pageants, complete with a plot of their own and with a motif chosen to be in direct contrast to the mood of the feature picture. The Jack Partington concept in San Francisco was to build the stage show around the personality of a showman-bandleader, who would present the acts in front of the orchestra on the stage. The Loew houses, under the guidance of Major Bowes, Chester Hale, and Louis K. Sidney, went in for an elaborate vaudeville policy with "name" attractions bolstered by a permanent orchestra and a line of chorus girls. And Fanchon & Marco, whose "Ideas" were to sweep the country in the late Twenties, used *all* these features in their units — one would be a miniature musical comedy, another a vaudeville revue, another a stage band unit, and another a pot-pourri of music and spectacle.

Chicago, busy in those days building up its reputation as a toddlin' town, was the nation's second liveliest theatrical center, and road shows would settle there for runs nearly as long as their New York originals. The legitimate theatre flourished on the windswept reaches of State, Randolph and Clark, but it was the movie palace that really put Chicago on the entertainment map in the Twenties. The Ascher Brothers chain with its 2,654-seat Sheridan; the Mark Brothers with their 3,978-seat flagship, the Marbo; and Lubliner & Trinz's 3,257-seat Belmont; together with each chain's hundreds of smaller houses, managed to do a profitable business in spite of the octopus-like competition of *the* great name in Chicago movie palaces, Balaban & Katz.

The first Balaban theatre, the Kedzie, had been opened in 1908 (that memorable year when the movies suddenly struck the nation like chain lightning). It had been established in a former shooting gallery on Chicago's Kedzie Avenue at Twelfth Street by Abe and Barney Balaban, some years before there was actually any Balaban & Katz organization. The Balabans might well have named it "The Family Theatre"; certainly, with sister Ida playing the piano, brother Max selling tickets, brother Dave taking them up, Abe and Barney cranking the projector, Papa acting as janitor, and Mama Balaban giving advice, it was a family enterprise on a grander scale than Sam and Rosa Rothapfel's backroom operation in Forest City, Pa.

One day the Balaban brothers met Sam Katz, a bright and ambitious young man about the South Side, who had gotten his start playing the piano in a nickelodeon in the afternoon when his mama thought he was off taking "classical" piano lessons. By the time Sam graduated from high school, he owned three nickelodeons of his own. The Balabans and Sam Katz knew they were meant for each other the moment they met, and it didn't take many evenings in Mama Balaban's kitchen to iron out the details of merger.

The first theatre built by the newly formed Balaban & Katz empire was the Central Park, opened in 1917 and seating 2,400 people. Designed by the new firm of George and C. W. Rapp, it was the first real movie palace in the Middle West, and in luxury and the lavishness of its presentations it rivaled anything in New York, including the Strand, Rialto, and Rivoli. Once on its way, B & K grew fast. By the mid-Twenties it was a multimillion-dollar operation with twenty-eight theatres in Chicago (including such super *palazzi* as the Chicago, the Uptown, the Tivoli, the Tower, the Paradise, McVicker's, the Oriental, the Roosevelt, and the Norshore—together seating more people than all the theatres on Broadway in New York) as well as over a hundred other theatres in other parts of Illinois, Indiana, Wisconsin, and Michigan.

Lochinvar from the windy city

Balaban & Katz's stage presentations had a flavor and an opulence all their own, just as their theatres did. The guiding genius behind the shows that played the B & K deluxe houses was a young man from New York named Frank Cambria who had come to Chicago to seek his fortune about the time the Balabans and Katz were beginning to make theirs. Cambria's idea was basically a simple one: each stage show would follow a certain unity of theme, with all the settings, songs, music, and costumes elaborating on a single motif that usually had no relation to the picture on the program with it.

Frank Cambria's Balaban & Katz presentations were different, lavish, and, above all, perfect in every detail. "I feel quite strongly," Cambria stated in an interview, "that there is place for but one plot in a performance, and when that performance takes place in a film house, that one must be the plot of the motion picture. Of two plots in a single performance, one is necessarily considered by the patron to be the better. The less meritorious one, which in the filmshow is that of the stage production, sustains the condemnation which the ticket buyer inevitably bestows. I simply remove the basis of comparison—the plot;

The "Little Chicago"

—the most exclusive theatre in the world

On the top floor of the Chicago Theatre building is the "Little Chicago," the private theatre of Balaban & Katz, and the only one of its kind in the world.

No admission price is charged to enter this theatre. The public has never seen it —doesn't even know it exists. It opens its doors only to the members of the Balaban & Katz organization. It is their work-shop.

The "LITTLE CHICAGO" seats only 250 people, yet it is as completely equipped as *any theatre in the land.* It has a wide, deep and complete stage, wings, drops, curtains, spot-lights, colored lights, moving-picture screen, dressing rooms and every facility and convenience known to the modern play-house.

The "LITTLE CHICAGO" is *a try-out theatre* where the executives of the Balaban & Katz organization, the theatrical experts, moving-picture experts, musical directors, light and scenic artists, gather to discuss and rehearse the specialties and pictures suggested for Balaban & Katz theatres. Nothing has a place on a program until it is passed upon and approved by this entire organization.

It is a very critical "audience." It is a *paid* audience, composed of men chosen from the very top of their professions. *Each is a recognized specialist.* And they use their combined brains and knowledge, and devote their lives to developing entertainment for *you.*

BALABAN & KATZ

Chicago | Tivoli | Riviera | Central Park

Roosevelt

this done, my presentations are judged upon their merits as entertainment."

Frank Cambria returned to New York after a merger with the Paramount–Famous Players theatres and Balaban & Katz created the renowned Publix Chain in 1926. Cambria (along with Jesse Crawford) was brought to New York as part of the spoils of war by the victorious Adolph Zukor. Or perhaps Cambria and Crawford came along in the caravan of Sam Katz, who arrived in New York at the same time to oversee the whole Publix operation, and considered *himself* the victor in the Paramount–Balaban & Katz–Publix machinations. One fact is indisputable, however: Chicago was the loser in the deal. It lost Jesse Crawford, one of the stellar attractions at the Chicago Theatre on State Street; and it lost Frank Cambria, whose fine Italian hand had created a whole *milieu* of movie-palace entertainment that had packed Chicagoans into Balaban & Katz theatres week after week for years. After 1926 Chicago began to decline from its once-proud position as "cinematic culture center of the world" (as the supremely pivotal Colonel McCormick had been pleased to call it), and though more and more lavish theatres continued to be built, things were never quite the same in "Chicagoland" again.

New York's Rivoli Theatre was extensively remodeled as a temporary showcase for Cambria while the Paramount was being built on the site of the old Putnam Building where Marcus Loew's offices once had been. A typical Cambria production was "Great Moments from Grand Opera," which opened at the Rivoli on March 2, 1926. *The Exhibitor's Herald* was ecstatic: "Somewhere, sometime, we've read about a young person named Lochinvar who came out of the West and stirred things up in a style peculiarly his own. That's ancient history, or fiction or something. Now we have history or fiction, or something repeating itself, and leave it to Broadway, that isn't fiction. Another Lochinvar in this case is Frank Cambria. His show put on at the Rivoli is real blown-in-the-bottle Balaban & Katz stuff, not the imitation with a brand as bogus as that found on what your b – – tl – gg – r whispers to you is of prewar vintage."

Frank Cambria (above) put Balaban & Katz on the map of Chicagoland with lush presentations like "Pearl of Baghdad" (upper right) and "Watteau Come to Life" (below).

The show opened on the Rivoli's brand new stage (replacing the old Roxy-inspired orchestra-and-soloist's platforms) with George Dayton introducing himself as the prologue from *Pagliacci* and explaining the idea in such a way that Rivoli patrons who had never been to the Met could tell what was happening. Then came the prayer scene from *Cavalleria Rusticana,* sung by a large mixed chorus, the women kneeling.

Dayton next introduced the "Miserere" from *Il Trovatore,* followed by the prison scene from *Faust.* The singing, again, was superb, but what brought bravos was the ascending of Marguerite to heaven, Little Eva fashion, and the descending of Faust to Hell, with the help of a trapdoor. The whole idea was disarmingly simple, and the secret lay in the absolute first-rate musical and vocal performance, and—above all—Cambria's vivid lighting, gorgeous costumes, and impressive (and costly) sets.

It is interesting to note that this same week a revel called *Just Girls* was playing at the Colony Theatre, a nearby Broadway movie house. It featured the Vanity Fair Orchestra made up of eleven girls, who played and sang Irving Berlin's "Always" and were followed by The Fourteen Dancing Dollies who "showed a bad lack of cadence in a time-step routine." Then came the Three Wainwright Sisters, "Queens of Melody" who sang "I'm Sittin' on Top of the World." "Ruby Keeler, a tap dancer," wrote the *Exhibitor's Herald* "might have stretched her time allowance and the audience would have been well satisfied." Then the entire company got together and gave a slam-bang finish, singing, "Bam, Bam Bammy Shore."

Croesus in concrete

Out on the West Coast creativity knew no bounds. In San Francisco Jack Partington (universally acknowledged as "one of the few real gentlemen in show business") was staging shows at the Granada Theatre that were colorful, lighthearted and bursting with mechanical innovation. Down in Los Angeles, two local institutions were making a mark that, like Partington's, would soon be felt all over the country: one was the team of Fanchon & Marco and the other was Sid Grauman, the beloved "Little Sunshine" of West Coast show business.

Sidney Patrick Grauman was born on the darlin' saint's day in Indianapolis in 1889. His father, David, was a minstrel-show man, and the Grauman family were virtually nomads during the early years of Sid's life. In 1898 he and his father followed the gold rush to Dawson City, Alaska, and it was here that Sid got his first taste of show business by putting on performances of somewhat dubious local talent for the benefit of miners whose luck had run out. Sid also sold month-old newspapers from the States for two dollars apiece for the benefit of the Graumans. Once he sold a paper to Wilson Mizner, the great and witty international con man, for fifty dollars, after agreeing to sell no more papers for an hour. Mizner hastily hired the back room of a saloon and charged news-hungry prospectors a dollar a a head to hear him read the paper aloud. When Mizner's "newscast" was finished, poor little Sid Grauman had no customers left for his two-dollar papers and Mizner, who should have been ashamed of himself, was three hundred dollars richer.

On their return to San Francisco, Sid persuaded his father that there was money to be made with the new fad, moving pictures, and together they

Sid Grauman, Hollywood's beloved "Little Sunshine."

opened the Unique Theatre on Market Street—a converted store equipped with eight hundred kitchen chairs, a giant operation as nickelodeons went, and a successful one. Alas, the Unique was soon destroyed in the Great Fire of 1906. But that didn't stop the Graumans. While the city was still burning, they put up a large tent and filled it with several hundred salvaged church pews. That night they opened for business with the slogan: "Nothing to Fall on You Except Canvas in Case of an Earthquake."

From canvas to the first all-concrete structure in downtown San Francisco was the next step for Sid and David Grauman. This was their Imperial Theatre, seating 1,800 people. By this time Sid had begun to feel his oats as a showman, and in a few months with the help of Jack Partington, the Imperial was making money with a highly successful policy of movies and elaborately presented vaudeville acts.

But San Francisco was not notable for being a "show town" and Sid Grauman was, first and foremost, a showman. So while managing an engagement of a revue he had produced called "Midnight in Frisco," at the Majestic Theatre in Los Angeles, he decided that here, amidst all the sunshine and show business, was the town for him. He persuaded his father to dispose of their San Francisco theatre holdings, and together they launched their most ambitious project to date. This was Los Angeles' first really de luxe movie palace, and they called it—with a typical pat on the pocketbook—Grauman's Million Dollar Theatre.

One of the attractions at the Million Dollar Theatre was its giant Mighty Wurlitzer pipe organ. Presiding at the console was Jesse Crawford whom Grauman had heard playing in Seattle and had brought to Los Angeles because he showed considerable promise.

When the Million Dollar Theatre was opened in 1917, a new kind of stage show was born: the Sid Grauman Prologue. Sid, who had been following the career of S. L. Rothapfel in the East with considerable interest and envy, realized that with his own talents as a showman, he could become the Roxy West of the Rockies (at slightly higher prices, yet). But instead of copying Roxy, he devised a different approach to movie-palace entertainment. Whereas Roxy believed in presenting a diversified program of high-toned musical acts that bore no relation to the theme of the feature picture, Sid built his Prologues around the picture, so that a Grauman show, movie and all, was a complete unit in mood, music, and pageantry.

Other theatres followed the Million Dollar. The Rialto (with a more modest policy of feature films and musical acts), opened the following year. On October 18, 1922, Grauman opened his first theatre in Hollywood, the Egyptian, one of the most sensational theatres architecturally ever built up to that time. The opening feature was that Prince of Premiere Performances (the Knickerbocker, the Rialto, the Rivoli, the Roxy-fied Capitol were all scalps on his belt), Douglas Fairbanks in *Robin Hood*.

The Egyptian's architectural wonders were not confined to the auditorium; pacing up and down on the parapet at the end of the great forecourt was a bearded Bedouin in a striped robe carrying a spear. He appeared before every performance (there were two a day, all seats reserved) and became one of the notable sights of Hollywood.

Grauman's next theatre was the Metropolitan, on the corner of Sixth and Hill streets in Los Angeles. It held 3,485 and was twice as big as the Egyptian in both seating capacity and magnificence. The style can best be described as Hispano-Persian, and when it opened on the night of January 26, 1923, with Gloria Swanson (also destined to do her part as a movie-palace opener) in *My American Wife*, the corner of Sixth and Hill looked like the piazza of Saint Peter's on Easter Sunday. Appearing in the prologue, as part of a cast of over a hundred, was Fred Waring and His Pennsylvanians, making their Los Angeles debut.

Grauman's most famous theatre was the Chinese. Located diagonally across Hollywood Boulevard from the Egyptian, it immediately became, for all time, the movie capital's town hall, mother

Grauman's advertising was almost as elaborate as the Grauman "prologues" which preceded every film.

church, forum, prado, and hall of fame. Any movie worth the film it was printed on was premiered at the Chinese. If the Roxy, in New York, was the Cathedral of the Motion Picture, the Chinese was its High Pagoda.

The Chinese Theatre opened on May 18, 1927, with Cecil B. DeMille's thirteen-reel *King of Kings* and a Biblical prologue that was, as one critic remarked, "the damndest thing this side of Oberammergau." According to legend, a few days before the opening, Grauman was watching masons smoothing the fresh-laid concrete blocks that paved the forecourt when he accidentally stepped back into a section that was still wet. The footprint gave him an idea. Under the kleig lights of opening night, Norma Talmadge became the first film star to leave hand- and footprints to the ages—a tribal rite that has been followed until finally there were no unpersonalized paving blocks left in the forecourt. It was whispered darkly over the less important tables at the Brown Derby that certain blocks were being freshly cemented over in dark of night to make room for immortalizing portions of the anatomy (foot- and hand prints being passé) of newer and brighter stars. Be that as it may, all the commemorative stones in the forecourt of the Chinese are scrawled with something nice about Sid Grauman. He died on March 5, 1950, at the age of seventy-one, and is buried in Forest Lawn. But his monument is on Hollywood Boulevard.

Grauman, a deeply religious man, credited his success to "the Man Upstairs," and often declared: "God does all my shows." The day after he died, Louella Parsons delivered a syndicated eulogy that concluded: ". . . it is comforting to imagine Sid close to the 'Man Upstairs,' placing the Grauman footprints in the Heavenly Forecourt, where an eternal record is kept of the men who do good in this world."

Sid Grauman Stages Fine Fashion Pageant, Prelude To "His House In Order"

Three of the beautiful mannequins in "Grauman's Pageant of Fashion" at his Million Dollar Theatre, Los Angeles, Cal.

Eighty-Eight Beautiful Models Display Newest Creations in Frocks and Frills While Spectacular Song and Dance Tabloid Production Makes Lovely Vehicle for Fair Posers

OUTDOING all his previous achievements, Sid Grauman made cinema theatre history the week of February 23 in Los Angeles, when he staged what he was pleased to style "The Pageant of Fashion."

Never before in the history of Los Angeles did the discriminating public of that city enjoy so elaborate a feast of beauty. With sixty lovely types of femininity as models and stage settings that rivalled the most daring creations of *Vogue* and *Vanity Fair*, Sid Grauman presented a Festival of Fashion in a class by itself.

Psychological insight was displayed by Sid Grauman in presenting his Fashion spectacle on the same program with "His House in Order," a Paramount-Artcraft picture featuring Elsie Ferguson. As in most of her vehicles, Elsie Ferguson wore some beautiful style creations, and so the atmosphere of the feature photoplay and the Fashion Pageant blended most harmoniously.

The Fashion Pageant was introduced with atmospheric film subtitles, which, in picturesque language, informed the audience that back in the days when the world was young, the Goddess of Fashion had derived her first inspirations from Mother Nature, but that today with myriads centuries intervening, the feminine devotees of the Goddess of Fashion were not to be found in mountain fastnesses or fair valleys, but in Tea Taverns, wherein might be heard melodious music and the footfalls of multitudes of whirling dancers.

Then the curtains parted, revealing a set, dazzling in its magnificence and representing the interior of a luxurious cabaret. The color scheme was gold and black. Attired as devils, brigands and other grotesques, Grauman's Symphony Orchestra played soft, intoxicating strains from a balcony, while below a bal masque was in full sway. The lights were dimmed even at first, and as they were made even dimmer, the dancers appeared as so many phantoms, till they finally whirled off the scene in darkness.

All lights suddenly flashed brilliantly, the orchestra struck up Drigo's Harlequin Serenade, interpreted by Irene Adams and Arnold Tamon, who suddenly emerged from the throng of gay masqueraders. Their costumes were exceptionally colorful and artistic and their dancing seemed to breathe the spirit of witchery and moonlight.

As their number concluded, down the grand escaliers which led from the balcony, there stepped a little chap just seven years old. His name was Jack Lloyd and he was a dashing, sparkling little chap. He sang with a jazzy intonation and a swagger that brought the house down; a little jazz song to the effect that with the present high cost of feminine apparel, a modern man would have been surely up against it in King Solomon's day. As he swung into the second chorus, a little maiden about four years of age, attired in harem costume appeared at the head of the great staircase and slowly descended. It was Mr. Grauman's idea to get away from the professional showgirl type. Each successive model seemed somewhat lovelier than the rest. The orchestra struck up "A Pretty Girl is Like a Melody" and an exquisitely beautiful brunette wearing a gown of silver, sleeveless and backless, walked down the staircase. The silvery gown was translucent, revealing subtle outlines of figure. Her skirt, raised daintily, revealed the latest Parisian hosiery of intricate design and filmy texture. From her high Spanish head-dress to her neat ankles, round one of which was twined a diamond bracelet, this girl was a poem. Every variety of feminine garb was displayed, from evening gowns to bathing suits. A tailored suit worn by the Queen of Spain at the somewhat recent races in Paris was reproduced, as well as a whole host of riding habits, sport suits, tennis garb, negligees and street apparel.

The firm of Myer Siegel and Company cooperated with Sid Grauman in presenting the pageant, which lasted about twenty-five minutes. Altogether, 88 people appeared in the act. From midweek indications, all previous attendance records at Grauman's Million Dollar Theatre were destined to be shattered.

A. J. and C. Garfunkel Take Site for Savannah Theatre

A. J. and C. Garfunkel have announced the purchase of property on the southeast corner of Abercorn and Congress streets, Savannah, Ga., for Arthur M. Lucas, owner of the Odeon and Folly theatres of that city. Mr. Lucas, according to announcement, is planning to build one of the largest and most modern motion picture theatres in the South.

The property has a frontage of 90 feet on Abercorn street and 180 feet on Congress. Some time ago Jake Wells, then owner of the Bijou Theatre of Savannah, took an option on the property with the view of building a motion picture and vaudeville theatre on the site, but this option was never exercised. The property is regarded as one of the most desirable locations in the business district, being only a half block from the main retail thoroughfare, Broughton street.

The ensemble of mannequins from Sid Grauman's "Pageant of Fashion" staged at his Million Dollar Theatre, Los Angeles, Cal. Sid can give some of our Eastern theatres cards and spades when it comes to putting on such a feature.

Jack Partington (above) originated the stage band idea and was first to install elevator platforms or "flying stages." Another of his discoveries was red-headed Paul Ash (right) shown here at the Granada Theatre in San Francisco surrounded by his aggressively admiring "Merry, Mad Musical Gang" and yards of shiny "dream cloth" drapes.

The rising young man of Market Street

To Jack Partington must go credit for all the flying and plunging orchestras, organists, and chorus girls that added such flexibility and surprise to stage presentations in all the best movie theatres. Partington's patented "Magic Flying Stages," as they were called, once and for all lifted the grand symphony orchestras and gold-crusted organ consoles from the dark obscurity of the orchestra pit, and gave performers onstage a magic carpet to sing, dance, soar, sink, and vanish on.

Jack Partington was born on the Isle of Man where his parents were vacationing. His father was an artist in Manchester, England, and when Jack was five the elder Partington saw an ad expounding the glories of the California coast. Off to Oakland he went, and soon sent for the family. Jack's first show venture was as a traveling pitchman up and down the Pacific Coast with a film showing the Twenty-Mule Team of Borax fame in action; Partington delivered the spiel while his companion cranked the projector.

In 1912 he had saved up enough money to open a nickelodeon of his own, on Hayes Street in San Francisco, but not enough to keep it going without working on the side as a private detective. His ambition and ideas attracted the attention of Sid Grauman, who was operating the Imperial

Theatre with vaudeville and movies. Grauman hired Partington to operate the Imperial for him when he left for Los Angeles. The Imperial was soon sold to Herman Wobber, the dean of West Coast exhibitors.

Wobber was at this time building a large new theatre called the Granada, and asked Partington to think up some innovations that would set it apart from the other theatres in the city. Partington, who had been dreaming of such an opportunity for years, didn't take long in coming up with a set of ideas that would make the Granada not only "different" in San Francisco but unique in the whole world. For it was the Granada in 1921 that became the first movie theatre in history to have a rising orchestra pit.

And that wasn't all; Partington also devised a way for the orchestra, once it had risen into spotlit splendor at stage level, to glide back across the footlights and rise again to newer heights up on the stage. To do this another Partington invention, the disappearing footlights, came into use; as the orchestra on its rolling "bandwagon" got ready to move onto the stage, the footlights obligingly flipped down in their trough out of harm's way. Once the orchestra was onstage, the footlights flipped back into position. Or, with the footlights still down, the platform of the orchestra elevator would become an extension of the stage proper, an apron that gave performers more intimacy with the audience. The footlights were gear-driven by an electric motor, but the bandwagon was maneuvered by hidden stagehands, tugging like barge mules until the orchestra was in position on the elevator section of the stage. In later years similar installations in the larger theatres had motorized band cars and as many as three separate stage elevators, so that while the orchestra was raised sixteen feet above stage level, a line of chorus girls could rise *ex machina* from the basement, and a trampoline act could vanish from sight in front of them.

At the Granada in San Francisco all this was done to carry out Partington's most unique contribution to the lively art of the stage show: the presentation of the "personality" bandleader.

Eddie Peabody at the Granada demonstrates a banjo-within-a-banjo as the girls wonder how they are going to get him down for the next show. Girl patriots on facing page have problems of their own.

Shows at the Granada were built around the stage band, and the stage band was built around one of the great phenomena of show business, Paul Ash.

Born Paul Ashenbrenner, he came from Milwaukee to San Francisco in 1918 where Partington saw him leading a pit orchestra in an Oakland vaudeville theatre. With his flaming mop of red hair and his flamboyant gestures, Ash soon had the audience watching him (or what they could see of him) instead of the show. To Partington, this meant only one thing: build a show around Paul Ash and let the audience enjoy this eccentric "Raggedy Andy of Music" to the fullest. So, up out of the pit at the Granada rose Paul Ash, and a new style of movie-palace entertainment was born.

Ash, and his "Merry Mad Musical Gang" set San Francisco on fire. Every week they would appear in a new setting and in some new and outrageous costume; one week Ash and the orchestra would all be Eskimos and perform in a giant igloo; another week, in kimonos and pigtails, they would go on a "trip to Chinatown"—or in jockeys' silks they would be off to the races. Partington provided Paul Ash with a troupe of regular performers who appeared with the stage band shows: George Dewey Washington, the dynamic Negro baritone; GoGo Delys, the personable chanteuse; plus Peggy Bernier, Milton Watson and the Eighteen Granada Sweethearts.

In 1925 Balaban & Katz raided Partington's lively repertory group and lured Paul Ash to Chicago with the promise of a theatre dedicated to his peculiar artistry—this was the Oriental on Randolph Street, a Near Eastern nightmare designed (obviously on a dare) by Rapp and Rapp. Partington, who had an exclusive contract with Ash, generously tore the contract up in order to free Ash for his big chance with B & K. At the Oriental he became the flappers' idol. But the flappers soon ceased to flap and turned into a new and noisier species—the shrieking, swooning female fan.

The "Paul Ash policy" was widely imitated after Ash's triumph at the Oriental. Rudy Vallee was resident personality at the Brooklyn Para-

mount two and a half years after the opening tour of duty by Ash himself. Rubinoff and his Violin held sway at the New York Paramount (following Ash's engagement there). Charles L. "Buddy" Rogers became a local celebrity in Pittsburgh at the Stanley Theatre (where Paul Ash never played). Jimmy Ellard was Omaha's darling at the Riviera, Vic Ince at the Alabama in Birmingham, Eddie Lowry in St. Louis, and Del Delbridge in Detroit were all as well known as the mayor in their respective towns. Don Albert and Dave Schooler, who played the Loew de luxe neighborhood circuit, had the Bronx, Queens, and Brooklyn eating out of their hands. At the Granada in San Francisco, diminutive Eddie Peabody and his banjo came flying up above the stage to take Ash's place. Peabody, with his wide-bottomed pants, his droll antics, and his Banjomanians, became as popular with San Franciscans as Paul Ash had been, and another career was launched on the Partington magic flying stages.

In 1928 Partington himself succumbed to the blandishments of Sam Katz and, after a few months' staging shows at the Paramount in Los Angeles, came to the paramount Paramount, the Publix flagship on Times Square in New York. Here he produced Publix Circuit units in rotation with John Murray Anderson, Boris Petroff, and Paul Oscard, under the guidance of Frank Cambria. Later, after S. L. Rothafel left the Roxy Theatre, Partington became producer of the Roxy stage shows for Fanchon & Marco who were operating the theatre, as well as President of Fanchon & Marco, Inc. At the Roxy Partington found a paradise of mechanical equipment to play with and a staff of musicians and artists (including a long return visit by his old protegé, Paul Ash) that helped him give full expression to all the ideas he had pioneered when the golden age of the movie palace was young.

Edna Wallace Hopper brought her art nouveau boudoir to the stage of the Granada in San Francisco to show the girls (a) how to primp, (b) how to take a bath in front of 3,000 people and (c) how to go nighty-night while Musette, the maid, marvels at her mile-a-minute delivery.

The "Idea" idea

"Mlle. Fanchon arrived in Atlanta yesterday aboard the *Crescent Limited* to supervise the rehearsal of the two carloads of Fanchon & Marco Sunkist Beauties who will appear in the opening show at the new Fox Theatre on Christmas Day. The presentation entitled 'Beach Nights' comes to the Fox direct from Philadelphia. Atlanta represents the forty-first week of Fanchon & Marco unit time which comprises a total of fifty-two weeks from coast to coast.

"No one seems to know just what Mlle. Fanchon's first name is. Ever since the days when she and her brother Marco composed a first-string vaudeville act, they have been known simply as Fanchon & Marco."

— The Atlanta *Journal*, December 22, 1929

Mlle. Fanchon—whose first name *was* Fanchon (her last name was Wolf) — was a long way from home when she stepped off the *Crescent Limited* in Atlanta. And a long way from the days when Mama Wolf would sound reveille at 5:00 A.M. in their home in Los Angeles so that little Fanchon and her brothers Mark (later latinized into Marco) and Rube could have a full day to practice their music and dancing. By the time they were in their teens, Fanchon and Marco were a success as a ballroom dance team in West Coast theatres. As the finale of their act, Marco would play the violin with Fanchon perched daintily on his shoulder as they exited picturesquely.

Around 1919 they began to produce revues, the first being called simply *The Fanchon & Marco Revue*. In 1921, spurred by local success, they produced a show called *Sunkist* and set out to tour the country. The tour almost came to grief in New York while playing the Globe Theatre on Broadway; an outfit named Rainbow Film Corporation filed suit against Fanchon & Marco for $100,000 damages, claiming that all fourteen scenes in the show "depict the conduct and operation of a motion picture company, and that there appears on the curtain and in these scenes the name 'Rainbow Film Company,' while one scene depicts the engaging of actresses in such manner as to be salacious, improper and immoral,

Marco

and that another scene describes the Rainbow Film Company to be a fake." The suit never got to court, as the Rainbow Film Corporation, the plaintiff, turned out to be something of a fake itself.

Once more back in California, Fanchon and Marco came up with an idea—an idea that was to spawn thousands of other "Ideas" for the next fifteen years. Why not, they reasoned, produce stage shows for movie theatres that would tour the country on a rotating basis and present, each week, a show far more lavish than a theatre could put together with local talent and local budget? This was in the days before the great theatre-chain

. . . Sunkist Beauties, friends, fans . . . and Fanchon.

circuits really got rolling—the nearest approach to the scheme had been the Roxy–Goldwyn Theatres wheel which suffered a permanent puncture when Joe Godsol got out of Goldwyn Pictures and Roxy got into radio. Marco got busy lining up theatres on the Coast, Miss Fanchon (she was known thus by all who worked with and for her—the Atlanta *Journal's* "Mlle." had been bestowed on her in an excess of Southern gallantry) hustled together the girls who had worked in the chorus of *Sunkist*, brother Rube Wolf assembled an orchestra, and the first Fanchon & Marco "Idea" was born.

An "Idea" in the Fanchon & Marco sense of the word was never just a vague mental image but a concrete and distinctly salable commodity. One Idea made a week, fifty-two Ideas made a year, and that was the calendar they went by all during the Twenties and into the Thirties.

The Ideas were for the masses. Regrettably, they edged out the symphony orchestras and corps de ballets in some of the larger theatres, but they brought live—and lively—entertainment into many theatres that would have otherwise ground out an endless stream of "kiddie revues," solos by choir-loft coloraturas, college-boy jazz bands, and smirking organists of more local notoriety than talent. In a Fanchon & Marco Idea, the master of

Give the Sunkist Beauties enough rope . . . and they would turn out to be fan dancers, mask dancers, ruffle dancers, shy butterflies or hippily haughty halberdiers.

ceremonies always got real chummy with the folks, the Sunkist Beauties looked lovely (from across the orchestra pit), and the settings and costumes were gaudy, professional and (for Wichita), dazzling. So were Idea titles like "Gyp Gyp Gypsy," "Eskimamas," "Fan Fantasy," "Skirts," "Eyes," "Dangerous Curves," and "Knice Knees" that sizzled on marquees around the circuit.

Fanchon & Marco hit the big time when they signed a contract with the fast-growing Fox Theatres chain. Then an arrangement with RKO Theatres as well as the Publix and the Poli New England chains expanded their playing time still more. In this way they could book talent at far less per week than a local theatre could (offering one or two weeks) because an F & M contract was good for a year or more of guaranteed work. Scenery and costume "depots" were established around the country, one in St. Louis and one in Niagara Falls, so that units originating on the West Coast would arrive in theatres in the Midwest and the East with all new scenery and fresh costumes. Prior to this, each unit had traveled with a complete duplicate set of scenery and costumes.

It was as easy to pick a bushel of Sunkist Beauties for a Fanchon & Marco Idea as it was to knock oranges off a tree in the San Fernando Valley. Every "ballet and military tap" studio in the United States was grinding out dancers, and most of them gravitated to Hollywood in the Twenties in hopes of getting into pictures; by the time the Vitaphone had swept on the scene in a flurry of all-talking, all-singing, all-dancing movies, the gravitation became an avalanche. One would-be Sunkist Beauty was Myrna Loy, who auditioned for Fanchon & Marco at the age of fifteen, fresh from her father's ranch in Montana.

"Myrna," recalled Miss Fanchon some years later, "was really a poor dancer, but her beauty was a strange, exotic type. The fact that she couldn't make good as a chorus girl was in her favor, for it showed marked individuality."

Harry Wenger
Photo

"Nile Nights" Idea started with a line of Red Hot Mummies.

Next, the Sunkist Beauties as fancy-dress Pharaoh's Daughters.

Miss Fanchon's somewhat enigmatic appraisal of Myrna's talent as a dancer was confirmed for all to see in 1929 when she participated in a markedly bovine *pas de deux* in Warner Brothers' all-talking-singing-dancing *Show of Shows*.

Others who rose to stardom after an F & M apprenticeship were Mark Plant, June Knight, Janet Gaynor (an extra girl in assorted tableaux), Doris Day. Lyda Roberti, who worked for a year as a chorus girl was allowed to sing "What Do You Do on a Dew, Dew, Dewy Day" one night in Los Angeles when one of the principals of the "Rainy Nights" Idea caught cold; Lyda was kept on stage for more than a half hour talking Polish, improvising and dancing, and, true to Hollywood legend, signed a picture contract the very next day.

The finale was right out of *Aida,* featuring (top to bottom) Lola and her Lotus Dance, Rube Wolf and his Tomb Twisters, and Amneris and her All-Girl Jazz Band. Plus the Sunkist Beauties again in daring costumes from some other Idea.

The Paramount Theatre in Los Angeles—Sid Grauman's old Metropolitan rechristened as "one of the Publix Theatres"—was the fountainhead of all the Fanchon & Marco Ideas. At the beginning of 1928, after six years of producing shows all over California, the contract with Fox West Coast Theatres was signed, expanding Fanchon & Marco's sphere of influence as far east as the Capitol Theatre in Salt Lake City. By January 1929 a total of thirty-one weeks' playing time was a reality when Fanchon & Marco's "Rolling On" Idea, featuring everything on wheels and rollers (including 30 Sunkist Beauties 30 on roller skates) rolled into Chicago. Then, during the year, as important way-stations were added in the form of opulent new Fox theatres in Detroit, St. Louis, Brooklyn, and Atlanta, the playing time across the continent reached fifty-two weeks, a full year of work for any healthy American girl who wanted to tap-dance, roller skate, ride a unicycle, balance on a ball, hang by her teeth and smile, smile, smile seven days a week at $38.50 per.

The Living Chandelier (left) from "Lights Out" Idea was good for the tummy muscles but tough on Baby Peggy, who looks as if she's holding the whole business on her poor little back. The Sunkist dryads (below) are out on a limb in the mistaken notion that only God can make a tree.

No. 4—CHESTER HALE

Sylphides, songbirds and salvation

When Roxy left the Capitol Theatre to establish a new diocese on the corner of Seventh Avenue and Fiftieth Street, Major Bowes finally had the theatre to himself. He created "The Capitol Family" to fill the old "Roxy's Gang" Sunday evening time on the Blue Network, and soon the austere and slightly pompous Major became—to his own surprise even more than anybody else's—as unctuous a paterfamilias of the air waves as one could wish for. But when it came to producing stage shows for the Capitol Theatre, he realized he needed a helping hand, and help came in the able person of Chester Hale, a handsome young dancing master who had performed in, and choreographed, a number of Broadway musical comedies.

The Chester Hale Girls were — until the Roxyettes pranced on the scene in 1927 — the best beloved group of young ladies in New York, a city famous for its appreciation of nifty hoofers. Each

A double dozen Chester Hale girls at the Capitol in New York rest their fine tail feathers in front of spectacular satin house curtain.

week Hale devised some new spectacle to display them in — the "Undersea Ballet," set to Debussy music, was one of his most widely copied diversions; it was a marvel of sea-green chiffon, rubber octopuses, scrim, magic-lantern waves, and Chester Hale Girls drifting through a canvas-and-gauze Davey Jones Locker on wires.

The Loew chain by this time owned the Capitol. When Metro-Goldwyn-Mayer pictures, a Loew subsidiary, was formed, Goldwyn Pictures' assets —including the Capitol Theatre—were part of the package. Loew's had begun a vast theatre-building expansion program around 1927, and the Capitol stage shows became the nucleus of a Loew Circuit, set up to play the increasing number of new de luxe Loew's theatres around the country. Actually, it was Joe Godsol's old scheme of a Goldwyn Theatres wheel, come to life again.

Big, bouncy Louis K. Sidney was brought in from a successful tour of duty at Loew's Aldine Theatre in Pittsburgh to direct the destinies of the Loew presentation at the Capitol, but now it was designed to fit more general tastes than those the Capitol had been catering to for so many years. A stage band policy was instituted, with Walter Roesner and his Capitolians onstage, sharing the musical honors with David Mendoza (and his associate conductor, Eugene Ormandy), and the Capitol Grand Orchestra in the pit.

The Chester Hale Girls were rechristened "The Chesterettes" — which was okay by Hale, but raised some question (as well as some eyebrows) concerning the physical endowments of the young ladies. The Capitol's old 1919 model Estey "grand concert organ" was given an overhauling with a lot of new Wurlitzer-type pipes and jazz effects, and Henry Murtaugh, one of the country's best theatre organists, was hired to preside at its console. The Capitol's stage was reconstructed at great expense, with the installation of Partington elevator sections, and a lift was placed under the orchestra pit to give Mendoza, Ormandy, and company more visibility.

In short, the Capitol was mending its fences against the attack on its flock that was sure to come when the Cathedral of The Motion Picture opened its doors only two blocks away.

But the most sweeping change of the L. K. Sidney regime was the introduction of variety acts on the Capitol's programs. The list of acts that played the Capitol between 1927 and 1935 (when the theatre finally succumbed to the Depression and went to an "all the show on the screen" policy) reads like the roster of a Sophie Tucker testimonial rally in Madison Square Garden. In the first three months of the new policy, alone, there were Nora Bayes, Anna Case, Morton Downey, Bobbie Arnst, Lester Allen, and Ben Bernie (all on one bill, along with a chorus of fifty-five, and a forty-girl ballet in a show called "The Spirit of Syncopation," in October 1927), and there were Van & Schenck, Paul Specht, Peggy English, Winnie Lightner, Georgie Tapps, Vincent Lopez, Rooney & Bent, John Charles Thomas, James Barton, Percy Grainger (who pantomimed "Country Gardens" while the Duo-Art Pianola did the playing), Albert Spalding, and Rubinoff (and *his* violin, too).

In the early Thirties, Arthur Knorr, a genial and experienced showman who had been the designer of Capitol stage settings in the Twenties, returned to assist in the production of the Capitol-Loew Unit shows. One of his most vivid recollections concerns the week of January 27, 1933, when Mary Garden made her long-heralded (and often postponed) appearance on the stage of the Capitol. Miss Garden, Knorr recalls, consented to attend the final dress rehearsal for the show, which was held on a Saturday morning. When she arrived at the theatre she was greeted by Major Bowes and personally conducted to her dressing room (hastily vacated by the orchestra's celebrated cellist, Yascha Bunchuck, and considered the best room in the house because of its accessibility to the stage). When The Mills Brothers, William Demarest and Estelle Collette, Peggy Taylor and Gary Leon, and The Maxellos had all done their turns, it came time for Mary Garden to appear for her part in the rehearsal. But there was not sign of her backstage.

"You'd better run catch Miss Garden," an usher whispered to Arthur Knorr. "She's left the theatre hopping mad and heading down Broadway."

Knorr dashed out and after a couple of blocks caught la Garden ankling toward Times Square as fast as her feet would carry her. "Why, Miss Garden," he puffed, running to keep up with her, "what's the matter?"

"I've never been so humiliated in all my life," was the stormy reply. "It's bad enough to ask an artist of *my* standing to use such a pig pen of a dressing room in the first place. But that's not the half of it. Mr. Knorr, do you know what I found when I opened the closet door to hang up my coat? *A pair of men's shoes!*"

Knorr thought fast, took a bead on the fancy bird he was pacing, and fired back: "Well—er—Miss Garden, you wouldn't have felt so bad if you had found them under your bed, would you?"

"Oh, aren't you sweet," gushed the diva. "Let's go back to the theatre!"

Later that year, when Major Bowes revealed that he had signed a contract with Sister Aimee Semple McPherson—the blonde star of Hollywood's Angelus Temple and the Four-Square Gospel circuit—to appear at the Capitol, anticipation ran high. Though seven years had passed since she "wandered in the wilderness" after her spectacular and more than somewhat dubious kidnaping, Sister Aimee was at the peak of her notoriety. She had even gotten herself obligingly involved in a juicy divorce wrangle with her sometime husband, David Hutton, and capturing her for her first personal appearance in a real theatre (the Angelus Temple didn't have a box office and therefore didn't count) was another medal on the Major's chest. In anticipation of a sell-out he arranged for a squad of policemen to be on riot duty in front of the Capitol the morning Sister Aimee arrived onstage to "lead her brothers and sisters of Broadway back to righteousness." But so few sinners turned out for the first services that the cops manned their barricades in lonely embarrassment.

Those who did attend were puzzled by the melange of sex and salvation that was thrown at them over the footlights. The act opened with the organ playing "Adeste Fidelis" (it was mid-September) as the curtains slowly parted to reveal Aimee in a skin-tight white satin gown complete with sequin cross, posed before a church window drop, holding a Bible. She was lovingly spotlit from on high. After delivering a rather autobiographical exhortation for fifteen minutes ("Oh, yes, brothers and sisters, golly gee and my, oh my—I'm so happy I'm saved—so happy I could struggle through life's stormy seas and burning deserts so I could, praise God, come here to Broadway, this citadel of sin, to bring you my message. It's a real glory, yes sirree, it's a glory . . .") she swooped to the footlights and led the Capitol congregation in a prayer for their wicked city. Then she exited to a timid sprinkle of hand clapping, the audience being divided on whether to applaud or say "amen." And the sin-cursed New Yorkers settled back to enjoy *The Solitaire Man* with Herbert Marshall and Mary Boland on the Capitol's screen.

Major Bowes had booked Sister Aimee for a week at $5,000 (with a split of the gross over $50,000), but he needn't have worried about having to split anything with her; she brought only $17,500 into the Capitol's collection plate for the whole week. (The theatre's average weekly take in those days usually ran to $40,000.) Her contract called for an appearance at the Fox Theatre in Washington (by then a Loew house) the following week, and an option for an extended tour of the other theatres in the circuit. But when Major Bowes tried to buy back the balance of the contract in an effort to save the Fox from the fiscal disaster that had just struck the Capitol (Aimee's engagement is still spoken of reverently in the Loew home office as the lousiest single week in history), Sister Aimee rolled her eyes to heaven, patted her blond finger wave, and said, "Oh, no, Brother Bowes. I'm *going* to Washington. It's God's will, and you know how He hates a welsher, hallelujah!"

They all played the Capitol: Babe Hoey's Peter Pan Orchestra (refugees from Capt. Hook's band?); Winsome Maxie Baer; the eternal unicyclist; Sister Aimee Semple McPherson (who wouldn't toe the mark); the unavoidable adagio team; and the owl and the pussycat — Mary Garden talking contracts with Major Bowes while Maestro (in silver frame) looks on apprehensively.

More stage-show personalities. Ruth Etting toured the Publix Circuit for more than ten cents a dance; Sharkey the Seal did it for fish; Ginger Rogers danced in the shows at the Paramounts (both New York and Brooklyn); Ken Murray, his trademark and girl friends, did time with Fanchon & Marco; L. K. Sidney signed Polly Moran for the Capitol in a big way; and there was always a dog act.

KEN MURRAY

A Roxyette never knew, from one week to the next, if she would be a dashing grenadier, a gay caballero, or worked to the bone, launched on a unicycle, roller skates or a big ball. And *that* took practice (right).

The silhouette effect (above) made giant shadows dance on the backdrop. Roxy finales were always special; in this one (right) the Roxyettes admire a group of jumping Jains as light bulbs sparkle.

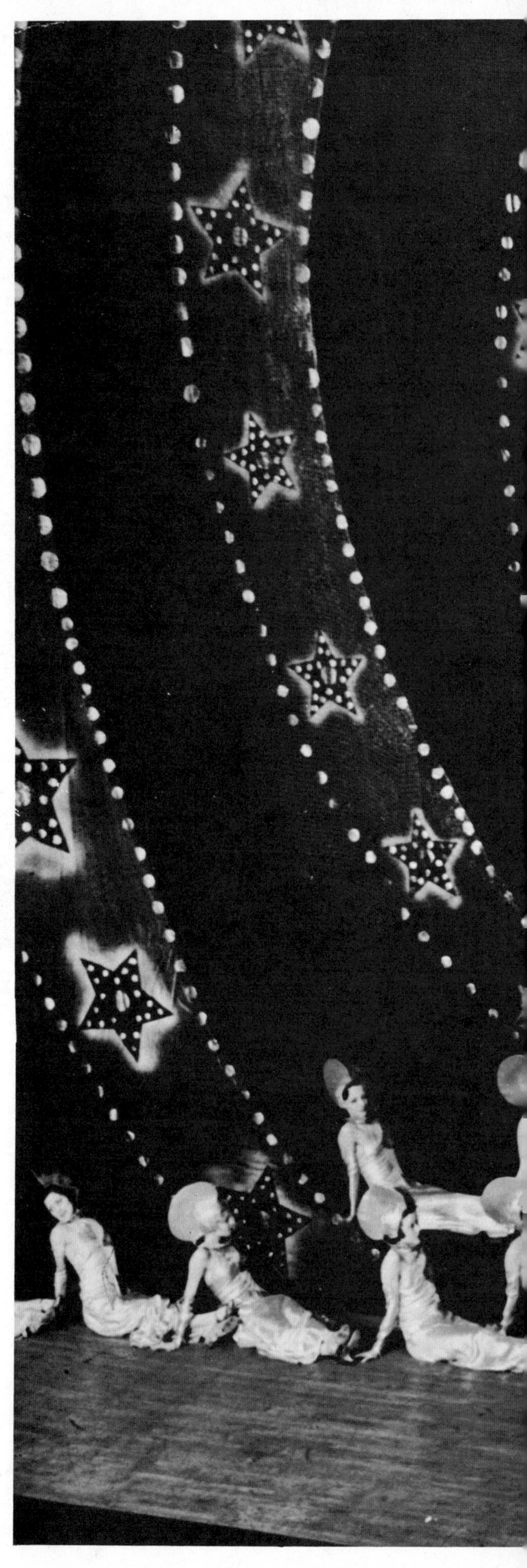

But no show at the Roxy was really perfect without Maria Gambarelli, assisted by a covey of sturdy sylphs from the corps de ballet. (See next page)

NA-SIB
NY 2.

"HEAR IT, CHEER IT . . . LOVE IT!"

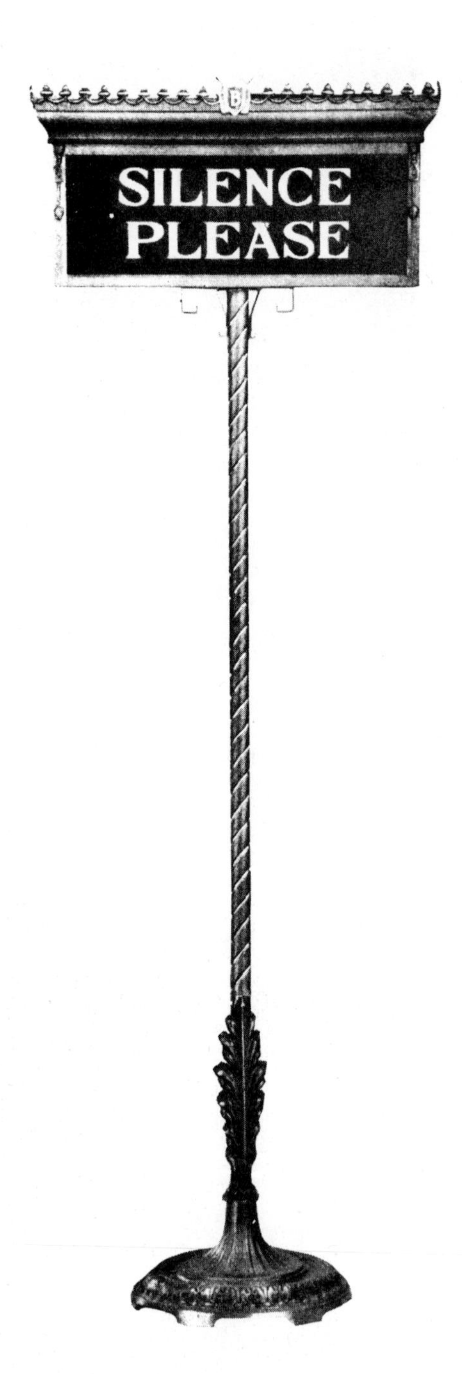

"We didn't need dialogue—we had faces! There aren't any faces anymore . . . there was a time when they had the eyes of the whole world; but that wasn't good enough for them. Oh, no! They had to have the ears of the world, too. So they opened their big mouths and out came talk, talk, talk."

— NORMA DESMOND
from the film, *Sunset Boulevard* — 1950

Contrary to popular legend, the first words spoken from the screen were not "Wait a minute—wait a minute, you ain't heard nothing yet." The story of talking pictures begins a long time before that night of October 6, 1927, when Al Jolson stunned the audience with the ad-lib heard 'round the world.

In 1887, two years before he introduced the first practical moving pictures with his Kinetoscope, Thomas Edison announced that he and his associate, William K. L. Dickson, were working on the development of "an instrument which would do for the eye what the phonograph does for the ear, and that by the combination of the two, all motion and sound could be recorded and reproduced simultaneously." Thomas Edison made it clear that talking pictures were not only a part of the original concept of the movies, but that he envisioned them chiefly as an accessory to his already established miracle, the phonograph.

Forty years were to pass before Edison's vision of talking motion pictures became a commercial success. Yet, on October 6, 1889, it became a practical reality when Dickson demonstrated the first model of the Kinetoscope to Edison with a short film in which he appeared and also said: "Good morning, Mr. Edison, I hope you are satisfied with the kineto-phonograph!"

Soon after the Kinetoscope (silent) became the rage, many establishments such as Raff and Gammon's Kinetoscope Arcade in New York were offering Edison *Kinetophone* sound subjects to patrons as well. Annie Oakley and a danseuse named Elsie Jones were among the first talking-picture stars in the world. By plugging the ears with stethoscope-like tubes, patrons could hear a somewhat synchronized accompaniment as The Little Sure-Shot fired away at some glass balls, or Miss Jones performed "a naughty buck dance." The sound was squeaked out by a cylinder phonograph installed in the base of the peep-show machine. In 1894 the Kinetoscope catalogue offered fifty-three subjects and noted that "we can furnish specially selected Musical Records for use on the Kinetophone, for nearly all the films on the foregoing list. Price, each, $1.50."

But the novelty of the Kinetophone soon faded, and in a few months it was withdrawn. When Edison finally perfected a system for projecting his moving pictures on a screen before a large audience, the ". . . two precious blond persons of the variety stage" who capered on the screen of Koster & Bials' Music Hall on the night of April 23, 1896, capered there in chaste silence.

Novelties don't stay novel very long in show business, and the fate of the Kinetophone was shared by virtually all the other talking picture devices that came along until the Vitaphone finally made the grade. For example, here are a few of the elaborately titled attempts at talking pictures that were made through the years:

THE PHOTOCINEMATOPHONE. In 1906 this pioneer sound-on-film system was developed in England by Eugene Augustin Lauste, a former Edison technician. Far ahead of his time, Lauste recorded his sound with a rudimentary microphone that translated the sound waves into a sound track of varying density on the edge of the film. In his projector a selenium cell was activated by light rays passing through the sound track. This is basically the same principle in universal use today, and it provided the answer to the synchronization bugaboo that was to finally be the downfall of the Vitaphone itself. But Lauste had no way of amplifying his sound for large audiences, and the Photocinematophone was a commercial failure.

THE CAMERAPHONE. In 1910 Edison tried talkies again, encouraged by the temporary success of the *Synchroscope*, a device imported from Germany by Papa Carl Laemmle of Universal Pictures. The Cameraphone coupled the projector to a phonograph behind the screen by means of a wire belt. The idea was startlingly simple; actually "talkies in reverse," it began with a phonograph record — then the recording artist was photographed pantomiming the words until a passable synchronization was achieved. Hollywood rediscovered the principle years later, and television has pounced on it even more recently.

HUMANUVA TALKING PICTURES. Another 1910 novelty was the Humanuva idea, which took the safe road of stationing live actors behind the screen to mouth the lines as they were pantomimed in the film. William Fox booked J. Frank Mackey's Humanuva version of Vitagraph's *Uncle Tom's Cabin* into his classy City Theatre on New York's Fourteenth Street where it caused something less than a sensation.

THE "NEW" KINETOPHONE. By 1913 Edison was at it again. On February 17 of that year he invited Martin Beck to go around with him to the four theatres in New York where his latest device was being introduced, to observe audience reaction. At the Colonial, the Alhambra, the Fifth Avenue, and the Union Square—all vaudeville houses—audiences saw and heard a scene from *Julius Caesar,* a short lecture by Edison explaining the device (actually an improved Cameraphone) that concluded with the Wizard of Orange, New Jersey, breaking a dish in perfect synchronization. A violinist, a singer, and a pianist were on the program, and the whole show wound up with a performance by a barking dog. *The New York Times* was somewhat cool toward the device, and observed next day that "for the present, at least, the prophecy that the 'talkies' will supplant grand opera or the legitimate drama seems fantastical."

WEBB'S ELECTRICAL PICTURES. On May 3, 1914, the Fulton Theatre in New York was the scene of an ambitious try at talking pictures. "One hour of Vaudeville, one hour of Opera, one hour of Minstrels," proclaimed the billboards, and *The New York Times* remarked on the process (which seems to have resembled Lauste's Photocinematophone of 1906) thus: "The sound of the voices in Mr. Webb's 'Electrical Pictures' is reproduced by a device whereby electrical vibrations are converted into natural tones. Both picture and voice are produced by the same apparatus and therefore the voice and action are synchronous."

THE PHONOKINEMA. In 1921 D. W. Griffith came up with this process to furnish sound accompaniment to his film, *Dream Street,* at Town Hall in New York. There were two hundred feet of actual dialogue in the film, the rest being musical background.

PHONOFILM. In 1923 Dr. Lee De Forest demonstrated that the great obstacle to amplification had been overcome. Using his "audion amplifier," which he had invented during World War I, in combination with a sound-on-film process almost identical to Lauste's, De Forest presented a series of short subjects that featured such leading vaudeville headliners of the day as Weber and Fields, Eddie Cantor, Phil Baker, Noble Sissle and Eubie Blake, Eva Puck, and Sammy White. The first showing was on April 15, 1923, at the Rivoli in New York, and later Phonofilm was exhibited in a hundred theatres in the United States and Canada.

THE VITAPHONE. At this same time the Bell Telephone Laboratories and their manufacturing division, Western Electric, were nearing the end of their experiments on a method for synchronizing films with sound on disks. By 1925 they had developed a special turntable capable of playing sixteen-inch disks at the same speed as today's LP recordings, with running time long enough to accommodate the sound for a whole reel of film. Bell's recordings were made with a microphone, not the horn-acoustical method used in commercial recording at the time, and the sound was surprisingly good. The synchronization (when everything worked) was good, too. Not being in the movie business, the Bell technicians tried to interest Hollywood in the process. But Hollywood turned a deaf ear to this newest novelty. Studios were making lots of money with silent pictures—so why make waves, sound or otherwise?

There was one company in Hollywood that wasn't making so much money. Unable to line up theatres to show their pictures (these were the days before Anti-Trust, and all the major producers held their own theatre chains), the Warner Brothers were deep in financial hot water. They decided to give the new talking-picture system a go. They christened it the Vitaphone, and in April, 1926, with the aid of Western Electric, set about perfecting it for theatre use. They providently signed exclusive contracts with Victor and Brunswick records and all their artists, and also made the same arrangement with the Metropolitan Opera Company. They leased the Manhattan Opera House on Thirty-fourth Street in New York for

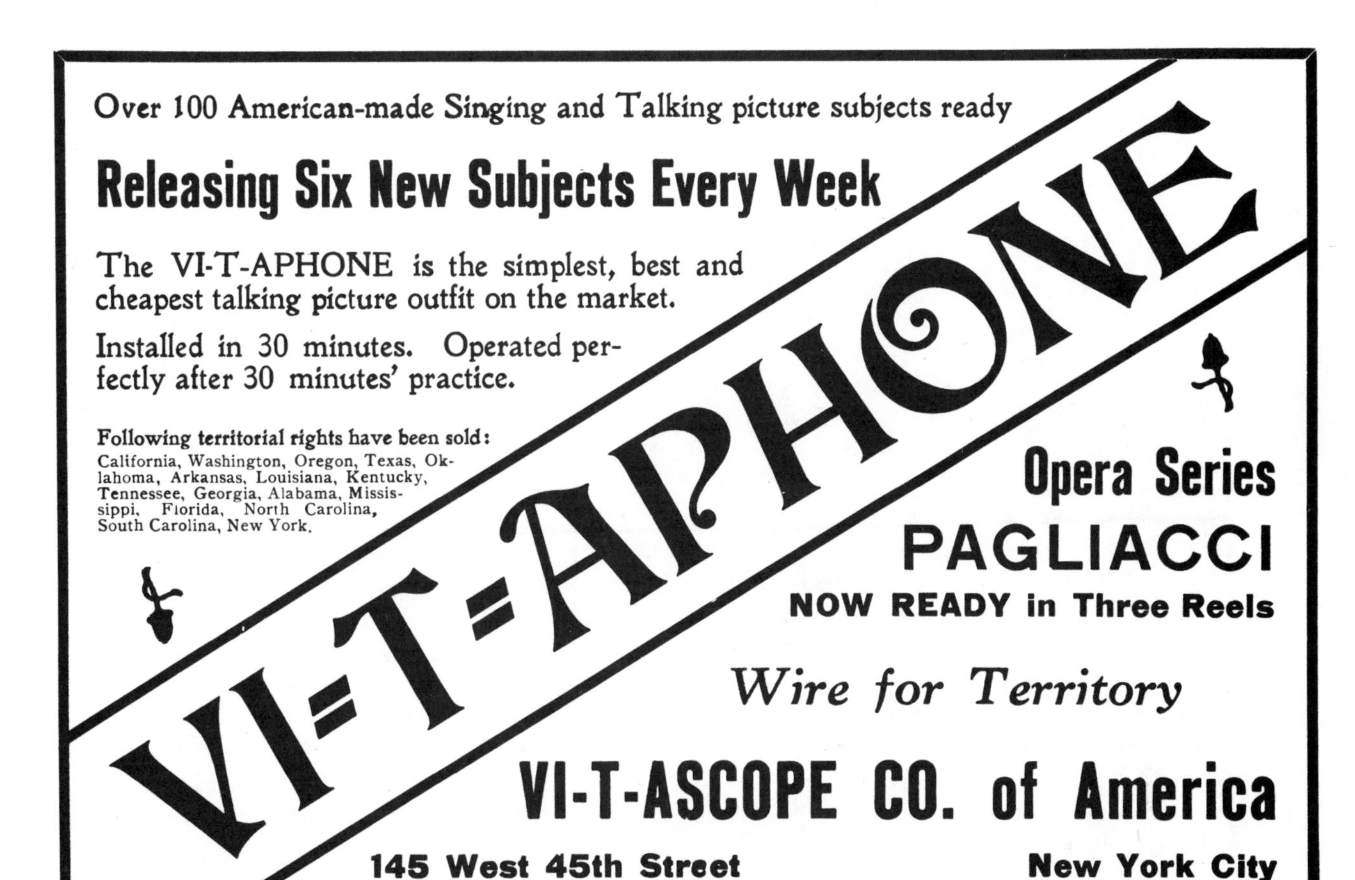

By 1913 there were all kinds of talkie systems, including a pre-Warner Vi-T-Aphone and the aptly named Mr. Blinkhorn's Vivaphone. The Excelsior Cabinet (right) made sound effects; the others tried to talk, but the synchronization bugaboo licked them

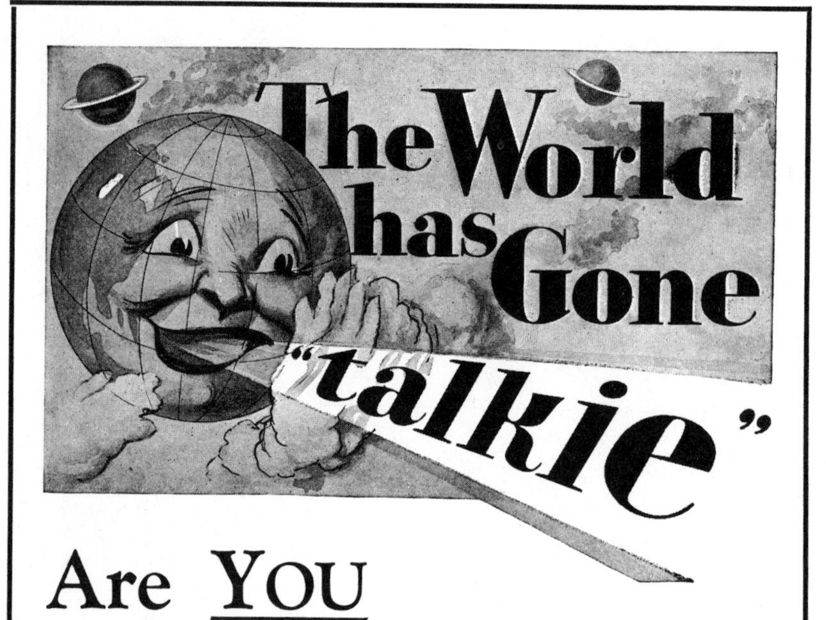

Are YOU
in line to supply the fans with *What They Want*?

INSTALL

SYNCRODISK

SYNCHRONIZED TURN TABLES

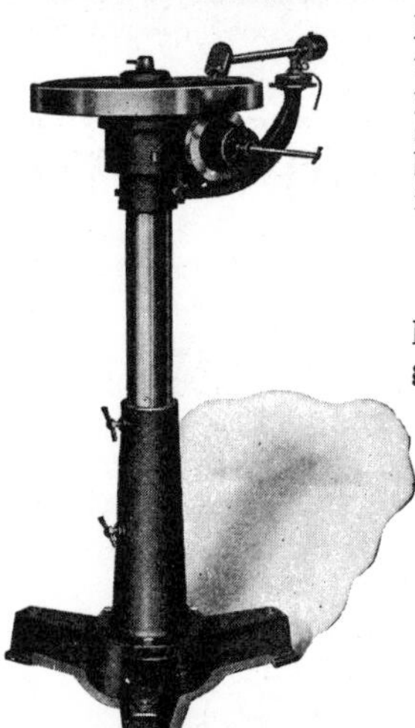

Perhaps you have been under the impression that it takes thousands of dollars to equip your theatre for talking pictures? If so, we suggest that you get in touch with us and learn all the details of Syncrodisk Synchronized Turn Tables.

***Complete for Only* $500**

Patrons are demanding talkies and they are going to the houses which show them. Get in line now at this low price.

HIGH POINTS

SPRING SUSPENSION: The Syncrodisk patent that spells steady, even motion.
METAL GUARD surrounding the disk has two pockets for needles.
NOT DRIVEN OFF INTERMITTENT.
INSTALLATION: Can be made by any operator and wire man. In one hour you are ready to run the standard 16" synchronized discs.
COMES COMPLETE with two pick-ups and change-over fader.

MORE THAN 300 now performing smoothly every day.
SUPPLIED with two Wright-DeCoster Speakers and one Amplifier for $750.00

Henry A. Lubé, European Representative

118 Blvd., Haussmann
Paris, France

Radio Motion Picture Co.
156 W. 44th St., New York

WEBER MACHINE CORPORATION

59 Rutter St., Rochester, N. Y.

further experimentation and recording. After a series of "by invitation" presentations, they were ready for the first demonstration of the Vitaphone before a paying audience on Broadway. The Piccadilly Theatre had been bought for the purpose, and renamed the Warner (not to be confused with the present Warner Theatre on Broadway which started its colorful history in 1914 as the Mark Strand).

There was no orchestra in the pit on the night of August 7, 1926, for the first time in the history of "de luxe" movie-theatre premieres. The show opened with a short talk by no less a personage than the "Czar," Will H. Hays.

The feature picture, *Don Juan,* had already been finished as a silent film when Warner Brothers took the plunge with the Vitaphone. The Barrymore epic represented a huge investment for the floundering Warners, so they decided to make it their first Vitaphone feature. The New York Philharmonic Orchestra was engaged, and the synchronized musical accompaniment for *Don Juan* was canned forthwith. There was no talking in *Don Juan,* though there were some synchronized saber clashes. The 127 kisses bestowed by The Great Profile upon Mary Astor and Estelle Taylor during the picture were—mercifully—non-sync.

This time there was quite a reaction—and a wildly mixed one. Will Hays wrote later that he said to himself, in the darkness of the theatre, while listening to himself talk to himself from the screen: "A new miracle has been wrought and I have had a part in it." One reviewer huffed that she thought "it was a bit of a comedown for Broadway audiences, used to 'live' orchestras in the pit, to have to put up with some Victrola records." Professor Michael I. Pupin of Columbia University sighed and declared: "No closer approach to resurrection has ever been made by science."

"I don't think," sniffed William Fox, "that there will ever be the much-dreamed-of talking pictures on a large scale. To have conversation would strain the eyesight and the sense of hearing at once, taking away the restfulness one gets from viewing pictures alone." Look who was talking!

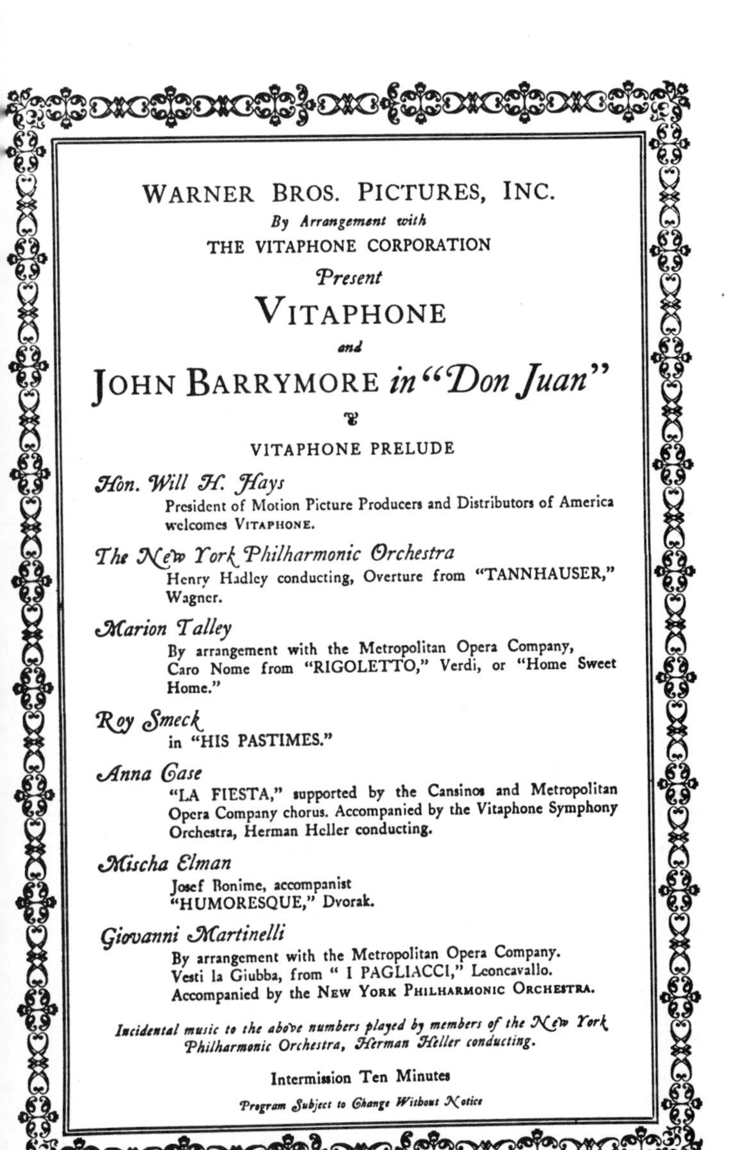

WARNER BROS. PICTURES, INC.

By Arrangement with

THE VITAPHONE CORPORATION

Present

VITAPHONE

and

JOHN BARRYMORE *in "Don Juan"*

VITAPHONE PRELUDE

Hon. Will H. Hays
President of Motion Picture Producers and Distributors of America welcomes VITAPHONE.

The New York Philharmonic Orchestra
Henry Hadley conducting, Overture from "TANNHAUSER," Wagner.

Marion Talley
By arrangement with the Metropolitan Opera Company, Caro Nome from "RIGOLETTO," Verdi, or "Home Sweet Home."

Roy Smeck
in "HIS PASTIMES."

Anna Case
"LA FIESTA," supported by the Cansinos and Metropolitan Opera Company chorus. Accompanied by the Vitaphone Symphony Orchestra, Herman Heller conducting.

Mischa Elman
Josef Bonime, accompanist
"HUMORESQUE," Dvorak.

Giovanni Martinelli
By arrangement with the Metropolitan Opera Company.
Vesti la Giubba, from " I PAGLIACCI," Leoncavallo.
Accompanied by the NEW YORK PHILHARMONIC ORCHESTRA.

Incidental music to the above numbers played by members of the New York Philharmonic Orchestra, Herman Heller conducting.

Intermission Ten Minutes

Program Subject to Change Without Notice

JOHN BARRYMORE *in "Don Juan"*

With Mary Astor
Adapted by Bess Meredyth
Directed by Alan Grosland

Musical score by Major Edward Bowes, David Mendoza and Dr. William Axt. Played on the Vitaphone by the New York Philharmonic Orchestra.

THE CAST

Characters	*Players*
DON JUAN	*John Barrymore*
Adriana Della Varnese	*Mary Astor*
Pedrillo	*Willard Louis*
Lucretia Borgia	*Estelle Taylor*
Rena (Adriana's maid)	*Helene Costello*
Maia (Lucretia's maid)	*Myrna Loy*
Beatrice	*Jane Winton*
Leandro	*John Roche*
Trusia	*June Marlowe*
Don Juan (5 years old)	*Yvonne Day*
Don Juan (10 years old)	*Phillipe de Lacy*
Hunchback	*John George*
Murderess of Jose	*Helene D'Algy*
Cesare Borgia	*Warner Oland*
Donati	*Montague Love*
Duke Della Varnese	*Josef Swickard*
Duke Margoni	*Lionel Braham*
Imperia	*Phyllis Haver*
Marquis Rinaldo	*Nigel de Brulier*
Marquise Rinaldo	*Hedda Hopper*

Licensed by New York State Motion Picture Commission—No. 145607.

Presentation directed by Herman Heller

The entire program produced and presented under the personal supervision of S. L. Warner

Warner Bros. Pictures, Inc., and The Vitaphone Corporation take this occasion to thank the Metropolitan Opera Company and the Victor Talking Machine Company for the great assistance rendered by them in connection with selecting the artists for this unusual musical program.

Marion Talley, Mischa Elman and Giovanni Martinelli appear by courtesy of the Victor Talking Machine Co.

The New York Philharmonic Orchestra records exclusively for the Brunswick-Balke-Collender Company, makers of Brunswick records. Mr. Elman was accompanied upon a Steinway piano.

But the Warner Brothers couldn't afford to listen to the scoffers; they had staked their very existence on the Vitaphone and there was no turning back. Two months after the opening of *Don Juan* they released Syd Chaplin playing "Old Bill" in *The Better 'Ole* at B. S. Moss' Colony Theatre, just up the street from the newly yclept Warner. This program opened on October 8, 1926, and like the first, was preceded by some Vitaphone vaudeville featuring Georgie Jessel and Al Jolson (strange film-fellows, as later events showed). On October 24, Warners' announced that the Vitaphone was available to all producers and exhibitors "of standing." No exclusive contracts were to be made, it was promised. Two weeks later Paramount, MGM, United Artists, and Pathé were dickering for a fifty per cent interest in the device. Fox wasn't heard from; he was off working feverishly on something called "Movietone."

MOVIETONE. While the other producers hurried to get aboard the Vitaphone bandwagon, William Fox set about securing the American rights to a German sound-on-film process called Tri-Ergon that seems to have antedated Dr. De Forest's Phonofilm system. Fox also paid a former associate of De Forest's, Theodore W. Case, a reputed one million dollars for the patents on still another sound-on-film process.

The Fox-Case development turned into the Movietone. It translated sound into lines of varying density along one side of the film, and besides being always in perfect synchronization, had the added advantage of being able to record sound outside the studio in the open air. The Vitaphone was tied to its turntables and recording styluses in the studio, but Movietone (which recorded inside the camera on the film as it was being shot), could operate anywhere a silent camera could.

Early in 1927, Fox began to release a series of Movietone shorts along the same lines as the Vitaphone subjects—filmed vaudeville acts. He also began to add musical scores to his silent feature films. *What Price Glory?* was the first. It opened on November 28, 1926, at the Harris Theatre (a legit house on Forty-second Street in New York), and on the bill along with the feature was a Movietone short of Raquel Meller, the Spanish singer. It was not until May 25, 1926, that Movietone released its first all-talking film-vaudeville program in connection with the premiere of *Seventh Heaven,* a silent picture with synchronized music. The program ran the gamut from Chic Sale to Ben Bernie to Gertrude Lawrence, and concluded with the most exciting thing Fox had turned out up to that time: shots of Lindbergh's famous takeoff—with sound recorded on the spot.

This single newsreel film marked the turning point in public acceptance of Movietone. It had been shown first in an extremely dramatic way: the second show at the Roxy on that memorable Saturday night, May 21, 1927, had been interrupted and the news of Lindbergh's safe arrival in Paris flashed on the screen. Then, while the audience was applauding, on came Fox's Movietone shots of the takeoff from Roosevelt Field. Six thousand, two hundred and fourteen breaths were held as the heavily overloaded little plane wobbled down the runway, its engine coughing in the cold dampness of the Long Island dawn. And when *The Spirit of St. Louis* finally cleared the threatening high-tension wires and soared off into the mist, the audience at the Roxy stood and cheered with the crowd at Roosevelt Field.

Throughout the summer of 1927 more theatres were wired for sound and more and more excitement was generated over the new idea of the talkies. Warner Brothers magnanimously agreed to permit theatres equipped for Vitaphone to also show Movietone features; a simple modification in the projectors made it possible. Though Vitaphone sound was on disks and Movietone was on the film itself, both depended on the Western Electric amplifiers and speaker horns to carry the sound to the theatre.

The Warner Brothers were working toward their big moment. In the spring they had bought the rights to Samson Raphaelson's stage hit, *The Jazz Singer,* and had contracted with its star, George Jessel, to re-create his role on the screen. But Jessel balked when he heard that the Warners expected him to *sing* as well as act in the picture. He had already appeared in a Vitaphone short, and knew that a lot of agonizing work and endless retakes were involved in an appearance before the Vitaphone camera. Jessel claimed his contract was for silent pictures only, and he wasn't working overtime for anybody without overtime pay. Exit Jessel, enter Jolson.

The film was shot as a standard silent picture, complete with subtitles, but it was Warners' idea to incorporate a few scenes along the lines of their successful little vaudeville shorts into the picture, in order to include the songs.

Jolson had had one fiasco in Hollywood back in 1923 when he walked out on a Griffith film called *Mammy's Boy.* Secure as long as he could perform behind the protection of his burnt-cork mask, he couldn't go through with the whiteface love scenes without clowning. This time it was different. He was a more mature performer, at the zenith of his popularity on the stage, and he was determined to not only make good in the movies now, but to enjoy himself doing it.

The picture opened at the Warner Theatre on October 6, 1927. It may have been a coincidence or it may have been planned by someone in Warners' publicity department with an ear for history; in any event, few people in the audience that night knew or cared that on that day exactly thirty-eight years before, William K. Dickson had spoken synchronized words to Edison in the first demonstration of the Kinetoscope.

Plans for the premiere had almost been called off. Sam Warner, the senior Warner brother and the real moving force behind the great Vitaphone gamble, had died suddenly in California the day before. None of the Warners were in the theatre that night, but their decision to go on with the premiere insured a lasting monument to their brother.

What the first night audience saw was a mediocre little film that tried to tie knots in the heartstrings. Visually static, it wouldn't have lasted two days as a silent. The sight of Warner Oland playing a cantor in a long gray beard, and a glimpse of an unidentifiable Myrna Loy in the chorus meant nothing in those pre-Charlie Chan, pre-Thin Man days. Even the Vitaphone accompaniment, ground out by the embarrassed New York Philharmonic, consisted of the same tired old fragments from Tschaikowsky, Grieg, and Brahms (plus some Jewish folk tunes in honor of the occasion) that had been coming out of orchestra pits for years. It was just the sort of "score" that the picture would have deserved if it had been shown as a silent in any movie palace in the country.

The evening opened with the usual Vitaphone variety bill. This time it was Eddie Peabody and his banjo, Ohman and Arden and their pianos, and Louis Silvers and his 107-piece New York Philharmonic. There was also a comedy sketch by William Demarest, but no invocation by Will Hays.

As *The Jazz Singer* unreeled, the first voice to be heard was not Jolson's. It belonged to Cantor Josef Rosenblatt, singing "Kol Nidre." But midway through the second reel came a scene in a sort of subterranean Childs' where everybody was pounding on the tables with wooden hammers in time to the music. Jolson, elbowing his way through a platter of ham and eggs, was persuaded (via subtitle) to be a sport and sing a song for the people. There was a moment of silence as the projectionist changed Vitaphone records. Then Jolson started in to sing "Dirty Hands, Dirty Face." He did a good job, and the extras expressed their enthusiasm by whacking the tables to splinters with their mallets. Jolson, forgetting for the moment that he was not on the runway of the Winter Garden, held up one hand and spoke the lines that will go down in show business history with those of Cohan, Barrymore, and West: "Wait a minute—wait a minute. You ain't heard nothing yet!"

Jolson went on singing. But the screen was never the same after that . . . somebody had *actually said some words* in the middle of a full-length feature film. And the somebody was the most popular entertainer of his day.

Alan Crosland, the director of *The Jazz Singer*, decided to leave Jolson's words on the disk and encouraged him to talk some more. Later in the picture, after Jolson had been singing "Blue Skies" to his mamaleh, Eugenie Besserer, he turned on the piano bench and started ad-libbing away to Miss Besserer's all-too-obvious amazement. She did little more than smile sweetly and murmur an occasional "Oh, that's nice," as Jolson rattled on about the fine clothes and the trip to Coney Island and the house in the Bronx that awaited her. But, after he had gone back to his singing and piano playing, there came the moment that is the real milestone of the piece. Cantor Oland, with beard bristling and eyes gleaming like some 1927 Jeremiah, burst into the room at the sound of the jazz music Jolson was making, and shouted:

"Stop!"

It was the first word spoken on the screen that ever advanced a plot.

The electric effect of *The Jazz Singer* was due not to the plot, not to the music, nor even to the still-new miracle of words coming out of an actor's mouth. The picture would have been "just another novelty" had it not been for Jolson. His lithe and bouncy *joie-de-vivre* came right out of the screen and hit audiences between the eyes and ears. Perhaps no other performer of the time could have served the cause of the talking picture so well as the nation's beloved Jolie.

The panic was on.

Theatres dashed into the race to install the Vitaphone. Fox began issuing regular weekly Movietone newsreels. And a flood of new devices appeared on the market. By 1929 there were ninety-four different systems advertised in the *Film Daily Yearbook*, including the Vitatone and the Moviephone.

•

In Hollywood the panic mounted to a fury. A ballad of the day summed it up:

"Lots of old-time stars are out of a job,
All they do is blubber and sob—
Ever since the movies learned to talk.
And a red-hot vamp who registers pash,
Sounds as if she's ordering hash—
Ever since the movies learned to talk."

The *Twentieth Century Limited*, westbound, was loaded to the uppers with actors from the "legitimate theatre," eager to harvest the manna in the new promised land. *They* could talk, and they knew it.

In Beverly Hills, murmurs of "How now, brown cow" came from every poolside. Declamations of "Lasca" and "Horatio at the Bridge" rent the air; cries for "Chloe" ricocheted off bathroom walls. And vocal coaches, elocution professors and expression teachers from all across America answered the call.

The mortality rate of film stars with angel faces and magpie voices was shocking after the first few months of the sound craze. The industry was in upheaval, and *Publix Opinion*, weekly confidential house organ of the Paramount-Publix theatre chain, wrote: "Just as the talking films will revolutionize the art of picture production, and many as yet undiscovered performers will be elevated to stardom, so will the new medium affect distribution, particularly in foreign countries. Some authorities point out that films made by 'Yankee' voices may not be acceptable in Dixie, and vice versa. At any rate, the universality of films is apparently at an end . . . or else they will make the whole world learn to speak the same language."

With Babel just around the corner, going to the movies became an impersonal affair. The Vitaphone might squawk its bravest . . . still, something was missing. Where was the magic moment when the balcony sitters could see the music stands in the sunken orchestra pit light up one by one during the last reel? Where was the carefree master of ceremonies with the wide-bottom pants and matching personality? Where were the girls with their blinding blondness, their beribboned tap shoes, their sorority *esprit*, their smile-darn-you-smiles? Where were the shows of yesteryear?

All on the screen. You hear it. You cheer it. You love it.

The stage moves to the screen

While Hollywood was busily turning itself inside out in behalf of the talkies, the exhibitors of the United States were a little less anxious to make the plunge. It cost money to convert a theatre to sound—upward of $10,000 to equip a modest-size theatre with the Western Electric system, plus Vitaphone and Movietone modifications to projectors, running another $3,000—and for a while there was a "wait and see" attitude. Adolph Zukor, with his far-flung empire of Publix Theatres, said, "It will take five years to permit us to perfect sound and screen devices, to achieve the required results. Many of the present players, who still may be popular then will have to take courses in elocution, and we will then be able to look at and listen to a motion picture without a subtitle or a spoken title." Mr. Zukor dreamed on for a while, but not for five years.

A few months after Zukor's "wait and see" pronouncement, his own chain of Publix Theatres began to feel the effects of talkies. Yet, in March, 1928, the stage-show circuit hit its all-time peak—a guaranteed thirty-seven weeks on the road for each unit, thirty-three weeks of playing time and four weeks of layoffs. Opening at the Olympia in New Haven at the top of the Publix wheel was John Murray Anderson's *"Knick-Knacks"*; Jack Partington's *"Hey Hey"* was jazzing it up at the New York Paramount; at the Metropolitan in Boston, Frank Cambria's *"Rio Romance"* starring Amata Grassi, the wife of Chester Hale. In addition to these and twenty-nine other units playing the Publix time, there was Mae Murray and her *"Merry Widow Revue"* touring a dozen of the super-deluxers, and Ruth Taylor, who was barnstorming through forty-six of the lesser Publix Theatres doing a single in behalf of her new silent picture, "*Gentlemen Prefer Blondes*."

But, while all seemed rosy as a bank of surprise pink baby spots, a little cloud had already appeared over the footlights. Only two weeks before, a complete "Vitaphone Manual" had been issued to all Publix managers. It told, among other things, how to program talking shorts in place of "some of the less punchy 'live' acts," how to plan advertising to introduce the miracle of the Vitaphone, and how to make the absolute most of "*The Jazz Singer*" when it finally got to your theatre. The case of the Rialto Theatre in Omaha was cited as a shining example of how to

Was it the Vitaphone – or was it Joe E. Brown or the igloo chill within that packed 'em in at the Warner Theatre in Hollywood in 1930? Only Warner Brothers knew.

exploit the "mammy" out of the Jolson epic. The manager simply persuaded the Union Pacific Railroad to offer excursion rates of nearly sixty per cent to everybody in a hundred and twenty neighboring towns who wanted to come to Omaha and witness the newest wonder of the world.

In a few months, *Publix Opinion*, Sam Katz's weekly missal to all the house managers in the chain, issued this front-page pronunciamento: "STUPENDOUS SOUND-SHOWS NOW PUT EVERY PUBLIX THEATRE IN DE LUXE CLASS." And Katz, with the eclat of a monarch bestowing wholesale knighthoods, declared that ". . . from this day forward, every Publix theatre is a De Luxe Theatre. Every Publix theatre manager is likewise a DE LUXE MANAGER."

By the spring of 1929 most of the Publix houses were advertising "The Vitaphone makes every seat in the house a front row seat." In all but the great *Super* De Luxe houses (New York, Chicago, Boston and a few others) the only live talent the newly "De Luxed" managers had left was the organist, the pit band, and a few stunned stagehands, all with local contracts yet to expire. In May of that year, Adolph Zukor, blessed with the kind of memory that makes an executive truly great, stated: "The silent picture, as an important part of the motion picture industry, is doomed by the advent of sound."

That wasn't all sound had doomed. In that same issue of *Publix Opinion*, the only acknowledgment that Publix units still made a few appointed rounds in widely scattered theatres was a note to the effect that when Dolores del Rio made her personal appearance at the Saenger Theatre in New Orleans (in connection with the premiere of her new Paramount picture, "*Evangeline*") there would be no unit show that week. The unit that should have played the Saenger on that date was canceled en route in Texas. "*The Desert Song*" via Vitaphone had moved in.

That week also marked the date when the Granada, San Francisco's inviolate temple of flying stage bands and spectacles, announced "A Revolutionary and Progressive Step in Entertainment—the STAGE MOVES to the SCREEN." The ad went on to say:

> Coming events cast their shadows before. The Granada does not propose to wait for the inevitable. The Granada is doing the thing now. The astonishing new idea of having every feature on the program in *picture with sound* is taking the place of everything that has gone before. The past is a dead issue. We have awakened from the lethargy and sameness of yesterday's entertainment and are springing into the sunlight of an achievement more dazzling and pleasing than anything which has gone before. The Granada Theatre is proud to lead the world's way in this exhilarating and happy entertainment idea, and proud as it can be to present a modern entertainment innovation in keeping with the ever-progressive demands of cosmopolitan San Francisco.

The Los Angeles Paramount—last of the Publix Pacific theatres to retain stage shows on a unit basis—followed the Granada into the ghost-world of "canned" shows the following week. That left Minneapolis as the western outpost of Publix time, since both Omaha and Des Moines had already gone "all sound." On May 25, 1929, the Rivoli and the Rialto in New York capitulated to the Vitaphone.

The veil had been rent. And all over the country the great plush house curtains with the gold tassels and the rhinestone butterflies began opening on nothing but picture sheets edged in black. One by one, the Publix units, footsore and pale beneath their Stein's Juvenile #3, shuffled into the Howard Theatre in Atlanta—the elephants' graveyard of Publix time. And even the Howard was not the same old jumping-off place; now it (along with dozens of others with once-proud local names like Riviera and Metropolitan and Granada and Capitol and Newman) had been "de-personalized" by the Home Office into just another Paramount.

After Atlanta, the units disbanded for what *Publix Opinion* called a "temporary reorganizational layoff"—the sister acts going back to their husbands, the adagio teams signing up for marathons in cheap ballrooms, the chorus girls opening up one-room dancing schools in the hinterlands, and the wonder dogs retiring to hot hotel rooms to dream of cool hillsides and frisky rabbits. Sure—they'd all be getting together in the fall when New York put out a call again.

But after Wall Street's grim performance on October 29, 1929, everybody was putting out calls of a different sort. The De Luxe Publix units were permanently disbanded. Only the great flagship houses—the Paramount in New York, the Metropolitan in Boston, the Chicago in Chicago, and a sprinkling of others in locations where pride and local custom demanded live stage shows along with the movies—kept up any semblance of the glory that once was Publix.

Fanchon & Marco began to feel the pinch, pared down the number of Sunkist Beauties in each Idea, and dusted off old scenery from the warehouses in St. Louis and Los Angeles for another trip around the fast-dwindling circuit. Loew's kept up a brave front for a while in the New York metropolitan area where the five Loew "wonder theatres" were still doing fair business with talkies and vaudeville shows "direct from the stage of the Capitol Theatre on Broadway." But elsewhere the once-busy Loew's circuit lost momentum and finally ground to a halt.

A chill settled over the loges that didn't come from the air conditioning, and going to the movies became a lonely experience though every seat might be filled. The noisy screen had filibustered its way to the top of the bill. The orchestra pits were empty, the dressing rooms began to fill up with old lobby displays and soft-drink cartons. Theatre managers, once a breed of busy impresarios bursting with fierce pride and a million ideas, turned from showmen into candy salesmen.

And there was popcorn in paradise.

Suddenly the screen discovered that it could produce spectacles with more people, more singing, more dancing, more music, more noise – and less taste – than any movie-palace stage show in the country. Talkies blared the death march of the live presentation and Hollywood worked the public into a frenzy of expectation with advertising like the traffic-stopping spectacular (see next page) that was unveiled on the façade of the Astor Theatre in New York on the night of August 14, 1929.

METRO
GOLDWYN MAYER
HOLLYWOOD
REVUE
25
STARS
CHORUS
OF
200

LUCKY
STRIKE

General Outdoor Adv Co
ENGEL & CO.
MEMBERS
NEW YORK STOCK EXCHANGE
MAIN OFFICE 120 BROADWAY
BERNARD JAFFE MANAGER
LUCKY STRI

APEDA
N.Y.

GOLDWYN-MAYER'S

TONIGHT SHOW

LLYWOOD

TONIGHT

EVUE

ASTOR THEATRE

25 STARS ★ CHORUS OF 200

ASTOR THEATRE

HOLLYWOOD REVUE

"IT'S TOASTED"

General Outdoor Adv Co

MCMP-18740

DECEMBER 23, 1932. AMUSEMENTS

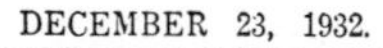
PHOTOPLAYS. PHOTOPLAYS.

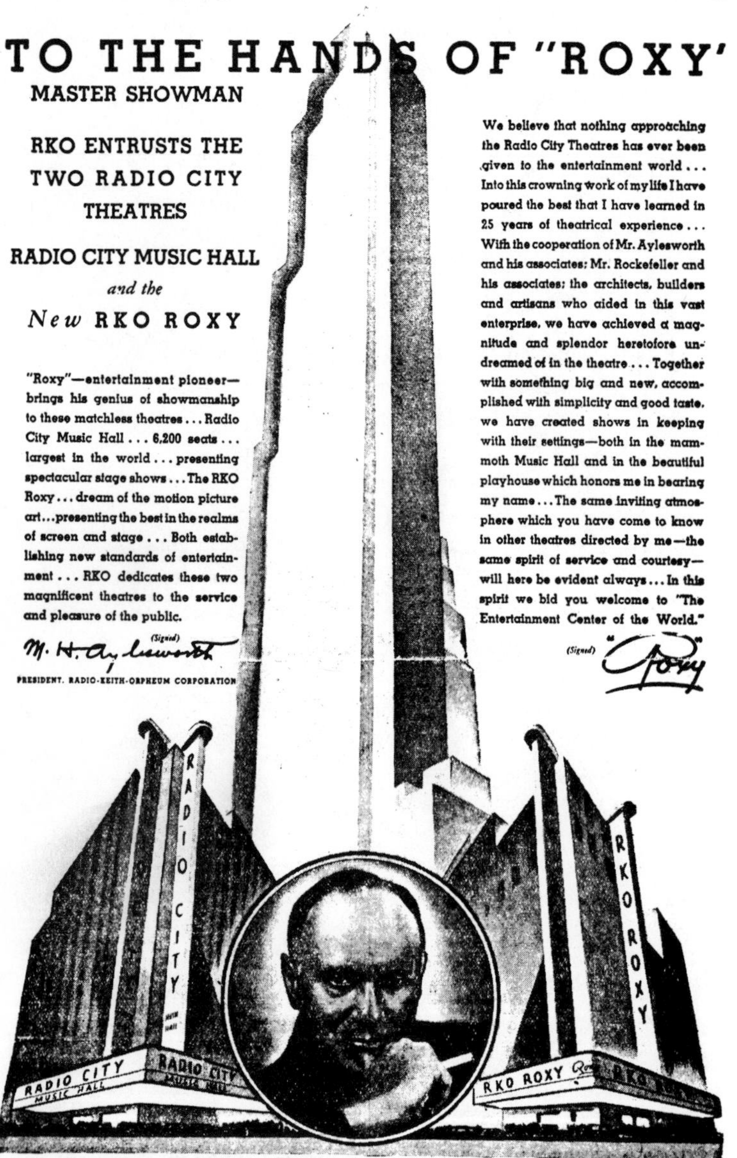

TO THE HANDS OF "ROXY"

MASTER SHOWMAN

RKO ENTRUSTS THE TWO RADIO CITY THEATRES

RADIO CITY MUSIC HALL

and the

New RKO ROXY

"Roxy"—entertainment pioneer—brings his genius of showmanship to these matchless theatres . . . Radio City Music Hall . . . 6,200 seats . . . largest in the world . . . presenting spectacular stage shows . . . The RKO Roxy . . . dream of the motion picture art . . . presenting the best in the realms of screen and stage . . . Both establishing new standards of entertainment . . . RKO dedicates these two magnificent theatres to the service and pleasure of the public.

(Signed) M. H. Aylesworth

PRESIDENT, RADIO-KEITH-ORPHEUM CORPORATION

We believe that nothing approaching the Radio City Theatres has ever been given to the entertainment world . . . Into this crowning work of my life I have poured the best that I have learned in 25 years of theatrical experience . . . With the cooperation of Mr. Aylesworth and his associates: Mr. Rockefeller and his associates: the architects, builders and artisans who aided in this vast enterprise, we have achieved a magnitude and splendor heretofore undreamed of in the theatre . . . Together with something big and new, accomplished with simplicity and good taste, we have created shows in keeping with their settings—both in the mammoth Music Hall and in the beautiful playhouse which honors me in bearing my name . . . The same inviting atmosphere which you have come to know in other theatres directed by me—the same spirit of service and courtesy—will here be evident always . . . In this spirit we bid you welcome to "The Entertainment Center of the World."

(Signed) Roxy

RADIO CITY MUSIC HALL

50th Street and 6th Avenue

OPENS TUESDAY EVENING, DECEMBER 27

Spectacular Stage Shows Twice Daily

BOX OFFICE NOW OPEN

Performances begin 2:15 and 8:15 . . . Doors open 1:30 and 7:30 . . All evenings and Saturday, Sunday, holiday matinees 99c, $1.50, $2.00, $2.50 . . . Monday to Friday matinees 75c, 99c, $1.50, $2.00 . . . Prices include tax . . . All seats reserved . . . Phone COlumbus 5-3030 . . . Mail orders filled on receipt . . . Special complete midnight show New Year's Eve.

New RKO ROXY THEATRE

49th Street and 6th Avenue

OPENS THURSDAY EVENING, DECEMBER 29

Fine Photoplays and Intimate Stage Attractions . . . World Premiere:

ANN HARDING and LESLIE HOWARD in "THE ANIMAL KINGDOM"

An RKO Radio Picture with William Gargan and Myrna Loy

From the Celebrated Play by Philip Barry

Continuous Daily from 10:30 A. M. to 1 A. M. . . . Popular Prices

Both Theatres Under Personal Direction of Roxy RADIO CITY ROCKEFELLER CENTER

THE END OF THE DREAM

"I would rather die with my boots on, to work until I drop. Death is just a big show in itself."

— S. L. ROTHAFEL, 1935

What happened to Roxy? In 1930 he had another vision, one that led him another block eastward along the same Fiftieth Street that had for so long been paved with the particular shade of antique gold he loved so much. The Rockefellers were building their Center, and in it the largest theatre in the world, the International Music Hall. This was to be all Roxy's, and so was a smaller theatre as well. The Radio City Music Hall (as it was finally named in deference to NBC, the Rockefeller's star tenant) would be the place where Roxy could devote all his energy to producing gorgeous music and spectacle . . . rejecting at last the movies that had made him, and which he increasingly ignored. The smaller theatre — the RKO Roxy — was to be given over to films and abbreviated pageantry.

Roxy pitched into the new project with his traditional energy and enthusiasm. But something happened, and the dream became a nightmare. "A spell of indigestion" in 1931 (his family never told him it was a heart attack) was the first warning of the weariness that was overtaking him. The face that had resembled a benign Il Duce became the mask of crumbling Caesar; in the final weeks before the Music Hall opened he was suffering so from acute prostatitis that a nurse was kept nearby at every rehearsal, on every tour through the huge modernistic theatre. Opening night, December 27, 1932, was a catastrophe; Roxy, in his fever to cram a life's ambitions onto a single stage in a single night (and Roxy, in his pain overlooking the little details of lighting, music, and taste that had always been his signature) had opened a Pandora's box of dullness and ostentation.

The show opened at eight-thirty with De Wolf Hopper, the Tuskegee Choir, and the Radio City Music Hall Ballet and Chorus in a curtain-raiser called "Minstrelsy," and plowed on through Russell Markert's Music Hall Roxyettes, Weber and Fields, Vera Schwarz, The Berry Brothers, The Wallendas, The Kikutas, Eddie and Ralph,

Harold Kreutzberg and Margaret Sande, Ray Bolger, Doctor Rockwell, Josie and Jules Walton, Dorothy Fields and Jimmy McHugh, Gertrude Niesen, a tab version of "*Carmen*" with Coe Glade, Desire De Frere, Aroldo Lindi and Patricia Bowman; Banto and Mann, Erno Rapee and the Music Hall Symphony, and Richard Leibert and C. A. J. Parmentier at the dual Wurlitzer consoles.

It was nearly 2:30 A.M. when the finale, "September 13, 1814"—Francis Scott Key and "The Star Spangled Banner" once more, and once more inspired (according to a press release) by a recent Rothafel shipboard reverie of dawn over the Atlantic — wound to a close with everybody (thousands?) rising out of the pit, rising out of the stage, revolving, singing, dancing, waving, bowing as the golden contour curtain lowered in swooping scallops to the stage. There was no movie.

Next day Roxy collapsed and was sent to the hospital. He may never have seen the reviews; they were very bad, and they might have been his death warrant. As it was, he was able to return to the Music Hall in a few months, only to find his hands tied by an "interim" management that now regarded him as an outsider; his shimmering spectacle trimmed to a charade, and movies on the screen. His RKO Roxy Theatre was being sued by the proprietors of the original Roxy Theatre over the use of his name — a name that was becoming strange to its owner—and was soon rechristened the Center Theatre.

The businessmen operating the Music Hall complained about Roxy's former extravagance in endless conferences, quite as if he were not present at all. It was a bewildering new experience for Roxy, all this talk about budgets and economy, but he assured them he would try. . . . In his heart, though, Roxy knew it was finished; there was no use staying on at the Music Hall where he wasn't loved . . . he was used to being loved. After all, wasn't he *Roxy,* and didn't he love everybody? So he left quietly one day and never came back.

When it became apparent that Leon Leonidoff, Russell Markert, Patricia Bowman, Charles Previn, Erno Rapee, Josef Littau, Florence and Hattie Rogge, even nurse Anne Bekerle—all "his" people whom he had brought with him from the Roxy Theatre—were going to stay with their jobs at the Music Hall instead of resigning with him in protest, Roxy's heart was broken: It was the dark of the Depression and jobs like those had vanished elsewhere, but in his grief Roxy didn't think of that.

For months he was too stunned to do anything but sit and wonder what had happened to the dream.

Then word came from Philadelphia that he was needed. The enormous Mastbaum Theatre—built just a few blocks too far from the center of the city to attract entertainment-shy Philadelphians—had closed in the face of terrible business, and its new owners wanted Roxy to come and work his old magic. But the magic that had brought glittering new life to so many theatres had lost its potency. The day of the Roxy-Mastbaum was brief, and the old magician wrapped his cloak about him and slipped away like a weary Merlin.

On January 13, 1936, Hope Williams, Roxy's secretary, entered his apartment in the Gotham Hotel in New York at 9:00 A.M., ready to work with him on his plans for a triumphal return to his beloved Roxy Theatre. There was a note on the desk asking her to call him at 9:30. Mrs. Williams stated, "I sent Johanna Rossi, Mrs. Rothafel's personal maid, to call him. In a little while she came back and said, 'I can't wake up Mr. Roxy.' Then I ran in and touched him. He had a smile on his face."

At his funeral, a squad of Marines fired a volley and a bugler blew taps. And as the echo floated over Linden Hills Cemetery in Brooklyn, someone whispered "Good night . . . pleasant dreams . . . God bless you."

•

Today the golden age of the movie palace has given way to an age of brass. Profit-conscious independent film makers are putting a strangle hold on theatre managers; theatre managers (most of whom are independent themselves, now) have reason to regret their haste in encouraging the antitrust suits that destroyed the great producer-owned chains like Paramount-Publix, Warner's, and Fox. It's every man for himself.

The movies, which got their start in storefront "theatres," have come full circle. The few new houses being built today are storefronts, too . . . places with seats and a screen and little else. Granted they are cooler, cleaner, smell better, and cost more to get into than the pioneer nickelodeons; they are also drab, antiseptic and earthbound. Projection may be flickerless, screens may assail the last frontiers of peripheral vision, sounds may come out of the walls, and seats may rock back like foam rubber dunking stools. But suppose the picture is bad? . . .

As for the dwindling number of genuine movie palaces that still open their doors, the going is getting tough. A few have had their faces lifted by uninspired interior decorators whose idea of cosmetic surgery is to smother every vestige of ornament, from proscenium to projection booth, in bolts of neutral-colored fiberglass. The graceful French curve of the New York Paramount's marquee has been supplanted by a frosted-glass trapezoid with plastic letters. An escalator now runs right up the middle of the Capitol's famous white marble stairs. Out in Hollywood, the foliated-gold interior of the Pantages Theatre resembles a yard-goods department and its seating has been drastically reduced to make the delegations who come to see the latest version of the Bible according to CinemaScope feel less lonely.

The clouds that once floated over a thousand balconies have drifted away for good. The machines broke years ago. One by one the stars have blinked out, their tiny bulbs blackened . . . dead stars in the cold outer space of grimy atmospheric ceilings.

Bowling alleys, supermarkets, garages, and apartment houses now loom where the once-proud Granadas, Strands, Rivolis, Tivolis, and Orientals stood.

•

In the summer of 1960 the Roxy Theatre vanished in a pile of rubble and dust and some shards of gold-leafed plaster of Paris. Fallow ground where only an office building could grow marked the realm where fantasy reigned, where romance and adventure flourished, where magic and charm united us all to worship at beauty's throne.

THE END

The death notice was brief, Gloria's visit poignant with old memories.

INDEX